AF443096

OSGi Service Platform, Release 2

OSGi Service Platform

Release 2
October 2001

Publisher

IOS Press
Nieuwe Hemweg 6B
1013 BG Amsterdam
The Netherlands
fax: +31 20 620 3419
e-mail: order@iospress.nl

Distributor in the UK and Ireland

IOS Press/Lavis Marketing
73 Lime Walk
Headington
Oxford OX3 7AD
England
fax: +44 1865 75 0079

Distributor in the USA and Canada

IOS Press, Inc.
5795-G Burke Centre Parkway
Burke, VA 22015
USA
fax: +1 703 323 3668
e-mail: iosbooks@iospress.com

Distributor in Germany, Austria and Switzerland

IOS Press/LSL.de
Gerichtsweg 28
D-04103 Leipzig
Germany
fax: +49 341 995 4255

Distributor in Japan

Ohmsha, Ltd.
3-1 Kanda Nishiki-cho
Chiyoda-ku, Tokyo 101-8460
Japan
fax: +81 3 3233 2426

LEGAL NOTICE

The publisher is not responsible for the use which might be made of the following information.

PRINTED IN THE NETHERLANDS

LEGAL TERMS AND CONDITIONS REGARDING SPECIFICATION

Implementation of certain elements of the Open Services Gateway Initiative (OSGi) Specification may be subject to third party intellectual property rights, including without limitation, patent rights (such a third party may or may not be a member of OSGi). OSGi is not responsible and shall not be held responsible in any manner for identifying or failing to identify any or all such third party intellectual property rights.

THE RECIPIENT ACKNOWLEDGES AND AGREES THAT THE SPECIFICATION IS PROVIDED "AS IS" AND WITH NO WARRANTIES WHATSOEVER, WHETHER EXPRESS, IMPLIED OR STATUTORY, INCLUDING, BUT NOT LIMITED TO ANY WARRANTY OF MERCHANTABILITY, NONINFRINGEMENT, FITNESS OF ANY PARTICULAR PURPOSE, OR ANY WARRANTY OTHERWISE ARISING OUT OF ANY PROPOSAL, SPECIFICATION, OR SAMPLE. THE RECIPIENT'S USE OF THE SPECIFICATION IS SOLELY AT THE RECIPIENT'S OWN RISK. THE RECIPIENT'S USE OF THE SPECIFICATION IS SUBJECT TO THE RECIPIENT'S OSGi MEMBER AGREEMENT, IN THE EVENT THAT THE RECIPIENT IS AN OSGi MEMBER.

IN NO EVENT SHALL OSGi BE LIABLE OR OBLIGATED TO THE RECIPIENT OR ANY THIRD PARTY IN ANY MANNER FOR ANY SPECIAL, NON-COMPENSATORY, CONSEQUENTIAL, INDIRECT, INCIDENTAL, STATUTORY OR PUNITIVE DAMAGES OF ANY KIND, INCLUDING, WITHOUT LIMITATION, LOST PROFITS AND LOST REVENUE, REGARDLESS OF THE FORM OF ACTION, WHETHER IN CONTRACT, TORT, NEGLIGENCE, STRICT PRODUCT LIABILITY, OR OTHERWISE, EVEN IF OSGi HAS BEEN INFORMED OF OR IS AWARE OF THE POSSIBILITY OF ANY SUCH DAMAGES IN ADVANCE.

THE LIMITATIONS SET FORTH ABOVE SHALL BE DEEMED TO APPLY TO THE MAXIMUM EXTENT PERMITTED BY APPLICABLE LAW AND NOTWITHSTANDING THE FAILURE OF THE ESSENTIAL PURPOSE OF ANY LIMITED REMEDIES AVAILABLE TO THE RECIPIENT. THE RECIPIENT ACKNOWLEDGES AND AGREES THAT THE RECIPIENT HAS FULLY CONSIDERED THE FOREGOING ALLOCATION OF RISK AND FINDS IT REASONABLE, AND THAT THE FOREGOING LIMITATIONS ARE AN ESSENTIAL BASIS OF THE BARGAIN BETWEEN THE RECIPIENT AND OSGi.

IF THE RECIPIENT USES THE SPECIFICATION, THE RECIPIENT AGREES TO ALL OF THE FOREGOING TERMS AND CONDITIONS. IF THE RECIPIENT DOES NOT AGREE TO THESE TERMS AND CONDITIONS, THE RECIPIENT SHOULD NOT USE THE SPECIFICATION AND SHOULD CONTACT OSGi IMMEDIATELY.

Trademarks

OSGi™ is a trademark, registered trademark, or service mark of The Open Services Gateway Initiative in the US and other countries. Java is a trademark, registered trademark, or service mark of Sun Microsystems, Inc. in the US and other countries. All other trademarks, registered trademarks, or service marks used in this document are the property of their respective owners and are hereby recognized.

Feedback

This specification can be downloaded from the OSGi web site:
http://www.osgi.org. Comments about this specification can be mailed to:
speccomments@mail.osgi.org

OSGi Member Companies

4DHomeNet, Inc.
Acunia
AMI-C
BMW
Cablevision Systems
Connected Systems, Inc.
Easenergy, Inc.
Elisa Communications Corp.
Ericsson
ETRI
Gatespace AB
IBM Corporation
inSilicon Corporation
ITP AS
KDD R&D Laboratories Inc.
Legend Computer System Ltd.
Mitsubishi Electric Corporation
Motorola, Inc.
Nokia Corporation
NTT
P&S Datacom Corporation
Patriot Scientific Corporation
ProSyst Software AG
Samsung Electronics Co., LTD
Schneider Electric SA
Sonera Corporation
Sprint Communications Company L.P.
Sun Microsystems
TAC AB
Telefonica I+D
Texas Instruments, Inc.
Toshiba Corporation
VDO Car Communications
Whirlpool Corporation

ABB Corporate Research Ltd.
Alpine Electronics Europe Gmbh
BellSouth Telecommunications
Bombardier Transportation
Coactive Networks
Deutsche Telekom
Echelon Corporation
Electricite de France
Espial Group, Inc.
France Telecom
Hewlett-Packard
Infomatec AG
Invensys Controls
Jentro AG
LANergy, Ltd.
Lucent Technologies
Metavector Technologies
Netpliance
Novell, Inc.
Oracle Corporation
Panasonic
Philips
Robert Bosch Gmbh
SBC Technology Resources, Inc.
Sharp Corporation
Sony Corporation
Symbol Technologies, Inc.
Telcordia Technologies
Telia Research
Tokyo Electric Power Company
Tridium, Inc.
Verizon
Wind River Systems

Board and Officers of OSGi

	Dan Bandera	Program Director & BLM for Client & OEM Technology *IBM Corporation*
President	John Barr	Director of Standards Realization, Corporate Offices, *Motorola, Inc.*
	Maurizio S. Beltrami	Technology Manager Interconnectivity, *Philips Consumer Electronics*
	Hans-Werner Bitzer M.A.	Head of Section Smart Home Products, *Deutsche Telekom AG*
Secretary	Robert Elliott	Director Central Electronics Systems, *Whirlpool Corporation*
Treasurer	Jeff Lund	Vice President, Business Development & Corporate Marketing, *Echelon Corporation*
	Stan Moyer	Strategic Research Program Manager, *Telcordia Technologies, Inc.*
	Hugo Österlund	Director of Operational Development, Corporate Technology, *Ericsson*
	François Rongere	CEO *Electricite de France* (EDF)
	Kjell Sullivan	Director, Home Server Development, *Nokia*
VP Marketing	Staffan Truvé	CEO, *Gatespace*
VP Americas	Bob Wambaugh, PhD	Interactive Television, Java Software, *Sun Microsystems*
VP Asia Pacific	Lawrence Chan	*Echelon Corporation*
CPEG chair	BJ Hargrave	Senior Software Engineer, *IBM Pervasive Computing*
Technical Officer/Editor	Peter Kriens	Managing Director, *aQute*
Executive Director	Dave Marples	Vice President, *Global Inventures, Inc.*
VP Europe and Asias	Thierry Nicolle	EMEA PVC Device Software Product Manager, *IBM Corporation*

Original Specification

The OSGi Service Platform, Release 2 specification has been reformatted to fit the size of a book. Errata that have been detected up to the March 2002 have been included in the book as well, changes to the specification are marked with change bars in the margin. Reformatting is done with great care and this book is intended to be a true copy of the content of the official OSGi specification except for the errata. However, the official reference specification is the PDF file that can be found on the OSGi web site: www.osgi.org.

Table of contents

Foreword

We've come a long way since the Open Services Gateway Initiative was created in March of 1999. From that early start of 15 member companies it has now grown to more than 80, with face to face meetings of over 150 people held three times a year all over the planet. In these times of increasing austerity it's testament to the importance that these member companies place on OSGi that we have such active membership.

When we started this work the original intention was to create a specification to allow services to be remotely deployed onto home network gateways – things like Set Top Boxes and DSL Modems. The first release of the specification explicitly addressed this market. This specification was extremely successful with many companies creating frameworks compatible with it.

Following the release of that spec many new companies joined the initiative and we started to experience pressure to increase the scope of the OSGi – taking it away from that narrow definition to become more of a *horizontal platform*, applicable in whole new environments that we never really foresaw in those early days, in markets as far apart as consumer electronics and automotive systems, security products and mobile phones.

Indeed, during the last year, in the face of all of this interest, we've had to go away and formally write down exactly what we're about to ensure all of this new blood didn't cause us to lose focus – the result of that was that we now have a clear, inclusive, statement of purpose; OSGi exists to create open specifications for the managed delivery of multiple services over wide-area networks to local networks and devices.

It's become apparent that OSGi principles are applicable in any environment where managed lifecycles, long uptimes and highly resilient, remotely managed platforms are requirements, and this breadth of applicability quite frankly took us a little by surprise considering the narrow focus we'd started from.

You'll see some of the early results of this expansion of focus in this document; new APIs such as Preferences and User Administration have been added and all of the existing capabilities have been enhanced. Despite this, one thing we've been very careful to do is to not alienate our early adopters, so you'll find very few incompatibilities between this specification and earlier ones.

We will continue expanding the specification and implementers can rest assured we'll give the same attention to backwards compatibility in the future.

As we said before, we're privileged to have world class people working on this initiative. This document is the product of the Core Platform Expert Group and those guys deserve a special pat on the back for their efforts.

Finally, we really do mean the Open in our name, so if the OSGi mission is important to you then please come and join us, and together we'll be able to make future releases even better...

So, here it is, the *OSGi Service Platform, Release 2*. Enjoy...

John Barr, *President OSGi*

1 Introduction

The Open Services Gateway Initiative (OSGi) was founded in March 1999. Its mission is to create open specifications for the network delivery of managed services to local networks and devices. With over 80 member companies today, OSGi is the leading standard for the next-generation Internet services to homes, cars, small offices and other environments.

The OSGi service platform specification delivers an open, common architecture for service providers, developers, software vendors, gateway operators and equipment vendors to develop, deploy and manage services in a coordinated fashion. It enables an entirely new category of smart devices due to its flexible and managed deployment of services. The primary targets for the OSGi specifications are set top boxes, service gateways, cable modems, consumer electronics, PC's, industrial computers, cars and more. These OSGi enabled devices will enable service providers like Telcos, Cable Operators, Utilities, and others to deliver differentiated and value added services over their networks.

Release 1.0 of the OSGi Service Gateway specification contained a specification for a service framework. This framework provides an execution environment for electronically downloadable services, called bundles. Deployed bundles are executed inside that framework and find a well-defined and protected environment. This environment includes a Java runtime and adds life cycle management, persistent data storage, version management and a service registry.

Services are Java objects implementing a concisely defined interface. The powerful OSGi framework registry is used to exchange services between bundles in a secure and controlled manner. Through this registry, bundles may provide services to other bundles as well as use services from other bundles. The registry is fully security protected, allowing the operator full control over the platform. Release 1.0 of the OSGi Service Gateway specification included the Framework and three basic service specifications: logging, a web server and device access.

This is the second release of the OSGi service platform specification developed by representatives from OSGi member companies. The first release has been available since May, 2000, and has set a standard for how managed services can be delivered over wide area networks to local area networks and devices.

This release represents the collective experience of implementors over the past year and the incorporation of new features to satisfy additional requirements. These new features have been developed by members of the OSGi Core Platform Expert Group, reviewed and approved by the entire expert group, and finally reviewed and approved by all of the OSGi member company representatives.

OSGi Service Platform Release 2 improves and extends the existing APIs. The modifications are backward compatible so that applications for the first release should run unmodified on release 2 Frameworks. And, the version management mechanisms allow bundles written for the new release to adapt to the old Framework implementations, if necessary.

The Framework specification is improved and clarified. Additionally, a number of features are added to simplify programming bundles. Security is strengthened and it is now possible to let a special management bundle fully define and control the security aspects of a bundle, in real time. Version management of bundles is hardened and interfaces are defined to make related information and control available to the management bundle. This release also includes a number of new service specifications.

1.1 Reader Level

This specification is written for the following audiences:

- Application developers
- Framework and system service developers (system developers)
- Architects

This specification assumes that the reader has at least one year of practical experience in writing Java programs. Experience with embedded systems and server environments is a plus. Application developers must be aware that the OSGi environment is significantly more dynamic than traditional desktop or even server environments.

System developers require a *very* deep understanding of Java. At least three years of Java coding experience in a system environment is recommended. A Framework implementation will use areas of Java that are not normally encountered in traditional applications. Detailed understanding is required of class loaders, garbage collection, Java 2 security, and Java native library loading.

Architects should focus on the introduction of each subject. This introduction contains a general overview of the subject, the requirements that influenced its design, and a short description of its operation as well as the entities that are used. The introductory sections require knowledge of Java concepts like classes and interfaces, but should not require coding experience.

Most of this specification is equally applicable to application developers and system developers. Items that are only relevant to system developers are marked appropriately, so that application developers can skip them.

1.2 Conventions and Terms

1.2.1 Typography

A fixed width, non-serif typeface (`sample`) indicates the term is a Java package, class, interface, or member name. Text written in this typeface is always related to coding.

Emphasis (*sample*) is used the first time an important concept is introduced.

1.2.2 Object Oriented Terminology

Concepts like classes, interfaces, objects, and services are distinct but subtly different. For example, "LogService" could mean an instance of the class LogService, could refer to the class LogService, or could indicate the functionality of the overall Log Service. Experts usually understand the meaning from the context, but this understanding requires mental effort. To highlight these subtle differences, the following conventions are used.

When the class is intended, its name is spelled exactly as in the Java source code and displayed in a fixed width typeface: for example the "`HttpService` class", "a method in `HttpContext`" or "a `javax.servlet.Servlet` object". A class name is fully qualified, like `javax.servlet.Servlet`, when the package is not obvious from the context nor is it in one of the well known java packages like `java.lang`, `java.io`, `java.util` and `java.net`. Otherwise, the package is omitted like in `String`.

Exception and permission classes are not followed by the word object. Readability is improved when the "object" suffix is avoided. For example, "to throw a `SecurityException`" and to "to have `FilePermission`" instead of "to have a `FilePermission` object".

Permissions can further be qualified with their actions. ServicePermission[GET,REGISTER] means a ServicePermission with the action GET and REGISTER.

When discussing functionality of a class rather than the implementation details, the class name is written as normal text. This convention is often used when discussing services. For example, "the User Admin service".

Some services have the word "Service" embedded in their class name. In those cases, the word "service" is only used once but is written with an upper case S. For example, "the Log Service performs".

Service objects are registered with the Framework. Registration consists of the service object, a set of properties, and a list of classes and interfaces implemented by this service object. The classes and interfaces are used for type safety *and* naming. Therefore it is said that a service object is registered *under* a class/interface. For example, "This service object is registered under PermissionAdmin."

1.2.3 Diagrams

The diagrams in this document illustrate the specification and are not normative. Their purpose is to provide a high-level overview on a single page. The following paragraphs describe the symbols and conventions used in these diagrams.

Classes or interfaces are depicted as rectangles, as in Figure 1. Interfaces are indicated with the qualifier <<interface>> as the first line. The name of the class/interface is indicated in bold when it is part of the specification. Implementation classes are sometimes shown to demonstrate a possible implementation. Implementations class names are shown in plain text.

Figure 1   *Class and interface symbol*

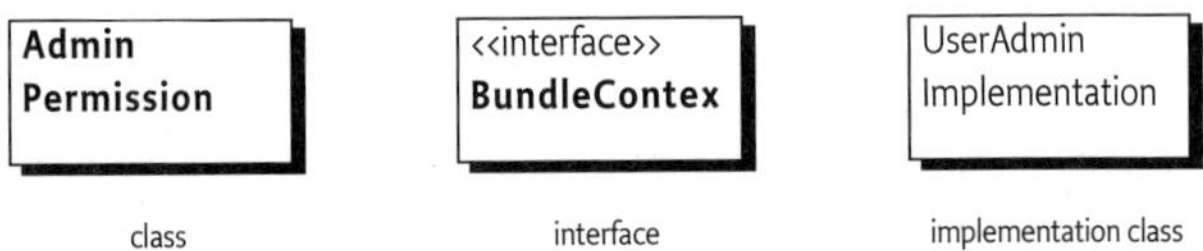

If an interface or class is used as a service object, it will have the qualifier <<service>> and a black triangle in the bottom right corner.

Figure 2   *Service symbol*

Inheritance (the extends or implements keyword in Java class definitions) is indicated with an arrow. Figure 3 shows that User implements or extends Role.

Figure 3   *Inheritance (implements or extends) symbol*

Relations are depicted with a line. The cardinality of the relation is
given explicitly when relevant. Figure 4 shows that each (1) Bundle-
Context object is related to 0 or more BundleListener objects, and
that each BundleListener object is related to a single BundleContext
object. Relations usually have some description associated with
them. This description should be read from left to right and top to
bottom, and includes the classes on both sides. For example: "A
BundleContext object delivers bundle events to 0 or more BundleL-
istener objects."

Figure 4   *Relations symbol*

Associations are depicted with a dashed line. Associations are
between classes, and an association can be placed on a relation. For
example, "every ServiceRegistration object has an associated ServiceReference object." This association does not have to be a hard
relationship, but could be derived in some way.

When a relationship is qualified by a name or an object, it is indicated by drawing a dotted line perpendicular to the relation and connecting this line to a class box or a description. Figure 5 shows that
the relationship between a UserAdmin class and a Role class is qualified by a name. Such an association usually is implemented with a
Dictionary object.

Figure 5   *Associations symbol*

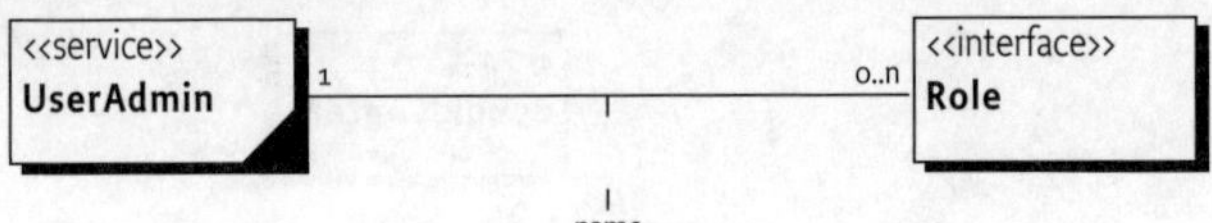

Bundles are entities that are visible in normal application programming. For example, when a bundle is stopped, all its services will be
unregistered. Therefore, the classes/interfaces that are grouped in
bundles are shown on a grey rectangle.

Figure 6   *Bundles*

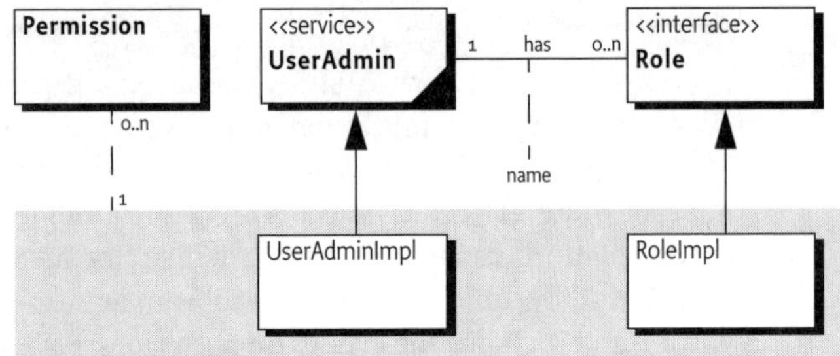

1.2.4 Key Words

This specification consistently uses the words *may*, *should*, and *must*.
Their meaning is well defined in [1] *Bradner, S., Key words for use in
RFCs to Indicate Requirement Levels*. A summary follows.

- *must* – An absolute requirement. Both the Framework implementation and bundles have obligations that are required to be ful-
  filled to conform to this specification.
- *should* – Recommended. It is strongly recommended to follow the
  description, but reasons may exist to deviate from this recommendation.
- *may* – Optional. Implementations must still be interoperable
  when these items are not implemented.

1.2.5 Term Definitions

- *Operator* - The operator is the organization that manages the OSGi
  Environment. It has full authority to modify the environment.
- *Service Platform* - A service platform is a Framework and a set of
  core service bundles.
- *OSGi environment* – The OSGi Framework is executed on sup-
  ported hardware platforms in a Java Runtime Environment (JRE).
  The combination of JRE and the OSGi Framework is called the
  OSGi environment. The OSGi environment is extended by
  installing bundles.

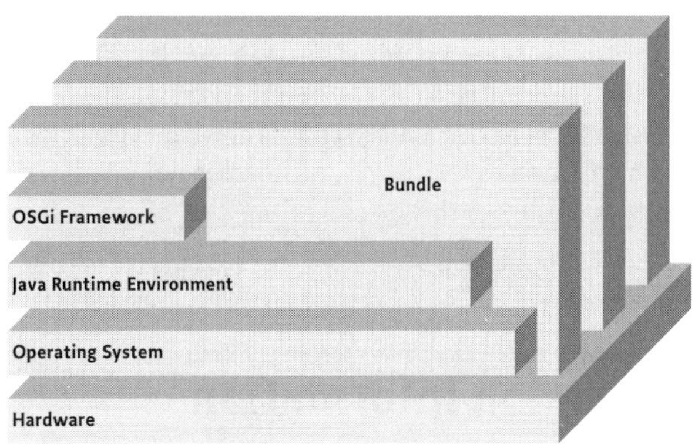

1.3 The Specification Process

Within the OSGi, specifications are developed by Expert Groups (
EG). If a member company wants to participate in an EG, it must
sign a Statement Of Work (SOW). The purpose of an SOW is to clar-
ify the legal status of the material discussed in the EG. An EG will
discuss material which already has Intellectual Property (IP) rights
associated with it, and may also generate new IP rights. The SOW, in
conjunction with the member agreement, clearly defines the rights
and obligations related to IP rights of the participants and other
OSGi members.

To initiate work on a specification, a member company first requests
for a proposal. This request is reviewed by the Market Requirement
Committee which can either submit it to the Technical Steering
Committee (TSC) or reject it. The TSC subsequently assigns the
request to an EG to be implemented.

The EG will draft a number of proposals that meet the requirements
from the request. Proposals usually contain Java code defining the
API and semantics of the services under consideration. When the EG
is satisfied with a proposal, it votes on it.

To assure that specifications can be implemented, reference implementations are created to implement the proposal. Test suites are
also developed, usually by a different member company, to verify
that the reference implementation (and future implementations by
OSGi member companies) fulfills the requirements of the specification. Reference implementations and test suites are *only* available to
member companies.

Specifications combine a number of proposals to form a single consistent document. The proposals are edited to form a consistent specification, which is voted on again by the EG. The specification is then submitted to all the member companies for review. During this review period, member companies must disclose any IP claims they have on the specification. After this period, the OSGi board of directors publishes the specification.

This Service Platform Release 2 specification was developed by the Core Platform Expert Group (CPEG).

1.4 Version Information

This document specifies OSGi Service Platform Release 2. This specification is backward compatible, unless indicated in the appropriate section, with the [2] *OSGi Service Gateway Specification 1.0* that was published by OSGi in May 2000. It specified the following components:

- Framework
- Log Service
- Http Service
- Device Access

This specification incorporates revisions of these components. Changes have been described in the description of each component.

New for this specification are the following components:

- Package Administration service
- Permission Administration service
- Configuration Administration service
- Preferences service
- User Administration service
- Metatyping
- Service Tracker utility (a version for Framework 1.0 is created as well)

Components in this specification have their own specification-version, independent of the OSGi Service Platform, Release 2 specification. The following table summarizes the packages and specification-versions for the different subjects.

A specification-version is needed in the declaration of the Import-Package or Export-Package manifest headers. Package versioning is described in *Sharing Packages* on page 20.

Item	Package(s)	Version
Framework	org.osgi.framework	1.1
Service Tracker	org.osgi.util.tracker	1.1
Permission Admin service	org.osgi.service.permissionadmin	1.0
Package Admin service	org.osgi.service.packageadmin	1.0
Log Service	org.osgi.service.log	1.1
Http Service	org.osgi.service.http	1.1
Device Access	org.osgi.service.device	1.1
Configuration Admin service	org.osgi.service.cm	1.0
Preferences Service	org.osgi.service.prefs	1.0
User Admin service	org.osgi.service.useradmin	1.0
Metatype	org.osgi.service.metatype	1.0

Table 1   *Packages and versions*

1.5 Compliance

OSGi environments that are designed to be compliant with this specification must provide:

- An implementation of the Framework as specified in this document.
- None of the other services or utility packages need to be implemented because they are all optional. However, when implementations of these services are provided, they should strictly follow the specifications in this document.

Conforming implementations must not add any new methods or fields to any of the classes or interfaces defined in the OSGi specification (though, they may add them to subclasses, or classes that implement the interfaces), nor may they add any new classes or packages to the org.osgi package tree.

1.6 References

[1]   *Bradner, S., Key words for use in RFCs to Indicate Requirement Levels* RFC2119, March 1997.

[2]   *OSGi Service Gateway Specification 1.0* www.osgi.org

2 Framework Specification

Version 1.1

2.1 Introduction

The Framework forms the core of the OSGi Service Platform Specification. It provides a general-purpose, secure, managed Java framework that supports the deployment of extensible and downloadable service applications known as *bundles*.

OSGi-compliant devices can download and install OSGi bundles, and remove them when they are no longer required. Installed bundles can register a number of services that can be shared with other bundles under strict control of the Framework.

The Framework manages the installation and update of bundles in an OSGi environment in a dynamic and scalable fashion, and manages the dependencies between bundles and services.

It provides the bundle developer with the resources necessary to take advantage of Java's platform independence and dynamic code-loading capability in order to easily develop, and deploy on a large scale, services for small-memory devices.

Equally important, the Framework provides a concise and consistent programming model for Java bundle developers, simplifying the development and deployment of services by decoupling the service's specification (Java interface) from its implementations. This model allows bundle developers to bind to services solely from their interface specification. The selection of a specific implementation, optimized for a specific need or from a specific vendor, can thus be deferred to runtime.

A consistent programming model helps bundle developers cope with scalability issues – critical because the Framework is intended to run on a variety of devices whose differing hardware characteristics may affect many aspects of a service implementation. Consistent interfaces insure that the software components can be mixed and matched, and still result in stable systems.

As an example, a service developed to run on a high-end device could store data on a local hard drive. Conversely, on a diskless device, data would have to be stored non-locally. Application developers that use this service can develop their bundles using the defined service interface, without regard to which service implementation will be used when the bundle is deployed.

The Framework allows bundles to select an available implementation at runtime through the Framework service registry. Bundles register new services, receive notifications about the state of services, or look up existing services to adapt to the current capabilities of the device. This aspect of the Framework makes an installed bundle extensible after deployment: new bundles can be installed for added features, or existing bundles can be modified and updated without requiring the system to be restarted.

The Framework provides mechanisms to support this paradigm which aid the bundle developer with the practical aspects of writing extensible bundles. These mechanisms are designed to be simple so that developers can quickly achieve fluency with the programming model.

2.1.1 Entities

Figure 8 on page 15 provides an overview of the classes and interfaces used in the org.osgi.framework package. It shows the relationships between the different Framework entities. This diagram is for illustrative purposes only. It shows details that may be implemented in different ways.

Figure 8   *Class Diagram* org.osgi.framework

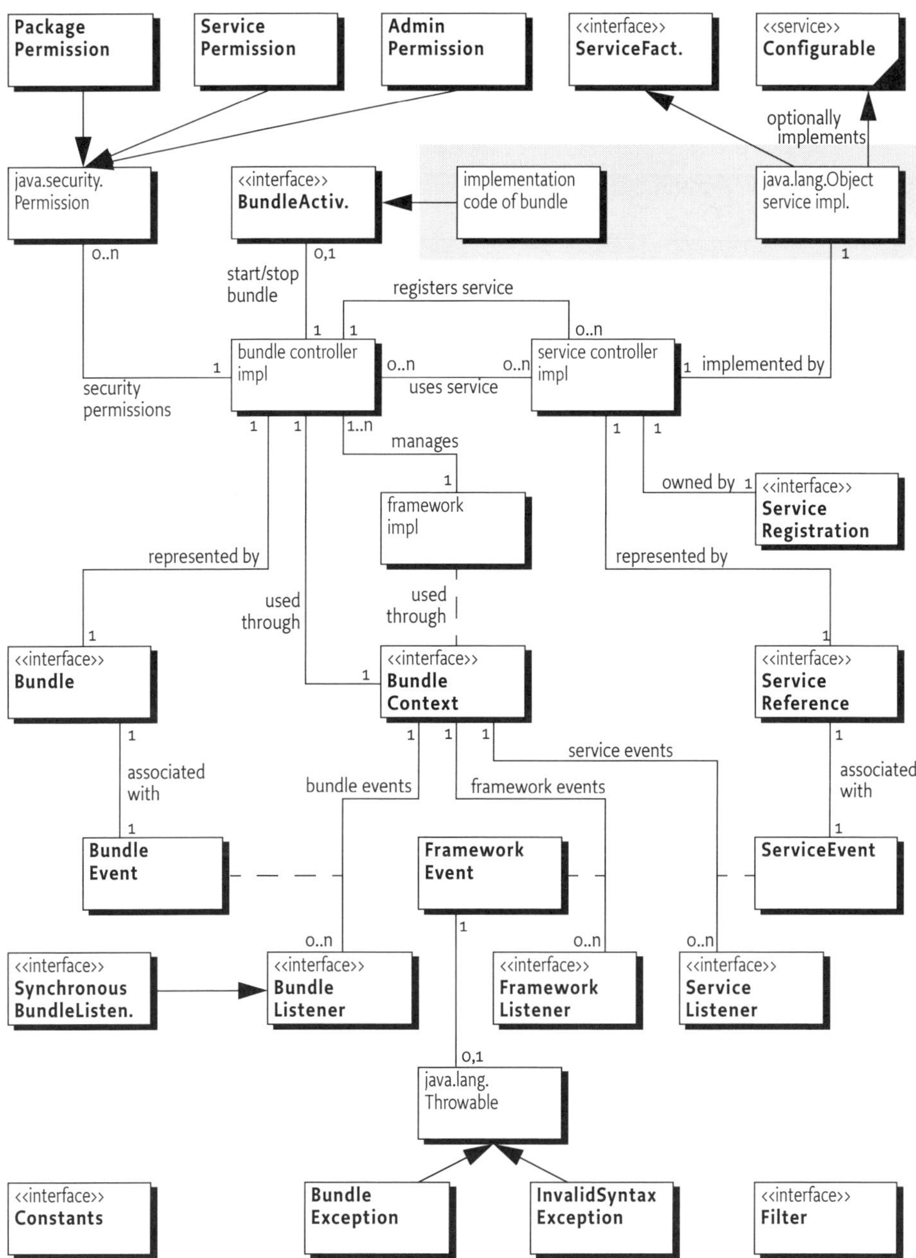

2.2 Bundles

In the OSGi environment, bundles are the only entities for deploying Java-based applications. A bundle comprises Java classes and other resources which together can provide functions to end-users and provide components to other bundles, called *services*. A bundle is deployed as a Java ARchive (JAR) file. JAR files are used to store applications and their resources in a standard ZIP-based Java file format.

A bundle is a JAR file that:

- Contains the resources to implement zero or more services. These resources may be class files for the Java programming language, as well as other data such as HTML files, help files, icons, and so on.
- Contains a manifest file describing the contents of the JAR file and providing information about the bundle. This file uses headers to specify parameters that the Framework needs in order to correctly install and activate a bundle.
- States dependencies on other resources, such as Java packages, that must be available to the bundle before it can run. The Framework must resolve these packages prior to starting a bundle. See *Sharing Packages* on page 20.
- Designates a special class in the bundle to act as Bundle Activator. The Framework must instantiate this class and invoke the start and stop methods to start or stop the bundle respectively. The bundle's implementation of the BundleActivator interface allows the bundle to initialize (for example, registering services) when started, and to perform cleanup operations when stopped.
- Can contain optional documentation in the OSGI-OPT directory of the JAR file or one of its sub-directories. Any information in this directory must not be required to run the bundle. It can, for example, be used to store the source code of a bundle. Management systems may remove this information to save storage space in the OSGi environment.

Once a bundle is started, its functionality is provided and services are exposed to other bundles installed in the OSGi environment.

2.2.1 The System Bundle

In addition to normal bundles, the Framework is itself represented as a bundle. The bundle representing the Framework is referred to as the *system bundle*.

Through the system bundle, the Framework may register services that may be used by other bundles. Examples of such services are the Package Admin and Permission Admin services.

The system bundle is listed in the set of installed bundles returned by BundleContext.getBundles(), although it differs from a normal bundle in the following ways:

- The system bundle is always assigned a bundle identifier of zero (0).
- The system bundle getLocation method returns the string: "System Bundle", as defined in the Constants interface.
- The system bundle cannot be lifecycle-managed like normal bundles. Its lifecycle methods must behave as follows:
  - *start* – This method does nothing because the system bundle is already started.
  - *stop* – Returns immediately and shuts down the Framework on another thread.
  - *update* – Stops and then restarts the Framework.
  - *uninstall* – The Framework must throw a BundleException indicating that the system bundle cannot be uninstalled.

  See *Framework Startup and Shutdown* on page 56 for more information about the starting and stopping of the Framework.
- The system bundle's Bundle.getHeaders method returns a Dictionary object with implementation-specific manifest headers. For example, the system bundle's manifest file should contain an Export-Package header declaring which packages are to be exported by the Framework (for example, org.osgi.framework).

2.2.2 Management Bundles

The OSGi Framework specification provides mechanisms but no policies. The implementation of the policies – for example, from what location a bundle is loaded – are provided by so-called *management bundles*. Management bundles have permissions to perform administrative functions on the Framework and provide the required policies.

Management bundles are normal bundles and should follow all the rules.

This specification does not define how an initial management bundle is installed in the Framework, because the installation process may differ for different deployment schemes. This pattern allows the operator to define the required policy in a simple and consistent way.

2.3   Manifest Headers

A bundle can carry descriptive information about itself in the manifest file that is contained in its JAR file under the name of META-INF/MANIFEST.MF.

The Framework defines OSGi manifest headers such as Export-Package and Bundle-Activator, which bundle developers can use to supply descriptive information about a bundle. Manifest headers must strictly follow the rules for manifest headers as defined in [11] *Manifest Format*.

All manifest headers are optional and any standard manifest headers not specified have no value by default (except for Bundle-Classpath that has dot (.) as default when it is not specified).

Consult the Constants interface for a list of standard manifest headers that may be declared in a bundle's manifest file.

A Framework implementation must:

- Process the main section of the manifest. Individual sections of the manifest may be ignored.
- Ignore undefined manifest headers. Additional manifest headers may be defined by the bundle developer as needed.
- Ignore unknown attributes on OSGi-defined manifest headers.

2.3.1   Retrieving Manifest Headers

The Bundle interface defines a method to return manifest header information: getHeaders(). This method returns a Dictionary object that contains the bundle's manifest headers and values as key/value pairs.

This method requires AdminPermission because some of the manifest header information may be sensitive: for example, the packages listed in the Export-Package header.

The getHeaders methods must continue to provide the manifest header information after the bundle enters the UNINSTALLED state.

2.3.2   Manifest Header Grammar

OSGi-defined manifest header values are declared using an augmented Backus-Naur Form (BNF) grammar similar to that used by [3] *The Standard for the Format of ARPA Internet Text Messages*. This augmented BNF is also specified by [4] *The Hypertext Transfer Protocol - HTTP/1.1* Section 2: Notational Conventions and Generic Grammar.

According to the BNF grammar, a manifest attribute value with spaces is parsed as a single word if the attribute is wrapped in double-quote marks.

2.4 The Bundle Namespace

This section addresses the issues related to classloading in the Framework and the details necessary to implement a Framework.

A *classloader* (ClassLoader object), loads classes into the Java Virtual Machine. When such classes refer to other classes or resources, they are found through the *same* classloader. This classloader may load the class itself or delegate the loading to another classloader. This approach effectively creates a namespace for classes: A class is uniquely identified by its fully qualified name and the classloader that created it. This implies that a class can be loaded multiple times from different classloaders.

2.4.1 Bundles and classloaders

Each bundle installed in the Framework that is resolved must have a classloader associated with it (Frameworks may have multiple classloaders per bundle). This classloader provides each bundle with its own namespace, to avoid name conflicts, and allows package sharing with other bundles. The bundle's classloader must find classes and resources in the bundle by searching the bundle's classpath as specified by the Bundle-Classpath header in the bundle's manifest. See *Bundle Classpath* on page 24 for more information on this header.

Bundles collaborate by sharing objects that are an instance of a mutually agreed class (or interface). This class must be loaded from the *same* classloader for both bundles, otherwise, using the shared object will result in a ClassCastException. Therefore, the Framework must ensure that all importers of a class in an exported package use the same classloader to load that class or interface.

For example, a bundle may register a service object under a class com.acme.C with the Framework service registry. It is crucial that the bundle that created the service object ("Bundle A") and the one retrieving it from the service registry ("Bundle B") share the same class com.acme.C of which the service object must be an instance. If Bundle A and Bundle B used different classloaders to load class com.acme.C, Bundle B's attempt to cast the service object to its version of class com.acme.C would result in a ClassCastException.

A bundle's classloader also sets the ProtectionDomain object for classes loaded from the bundle, as well as participates in requests to load native libraries selected by the Bundle-NativeCode manifest header. The classloader for a bundle is created during the process of resolving the bundle.

2.4.2 Sharing Packages

A bundle may offer to export all the classes and resources in a package by specifying the package name in the Export-Package header in its manifest. For each package offered for export, the Framework must choose one bundle that will be the provider of the classes and resources in that package to all bundles which import that package, or other bundles which offer to export the same package.

Selecting a single package among all the exporters ensures that all bundles share the same class and resource definitions. If a bundle does not participate in the sharing of a package – in other words, the bundle does not have an Export-Package or Import-Package manifest header referencing the package – then attempts by the bundle to load a class or resource from the package must not search the shared package. Only the system classpath and the bundle's jar file are searched for this package.

Figure 9   *Package and class sharing*

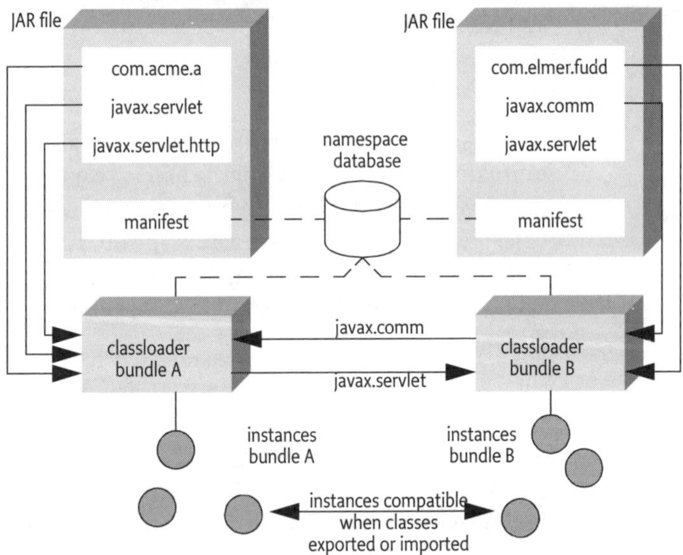

Package sharing for a bundle is established during the process of resolving the bundle. A bundle must only participate in sharing packages if the bundle can be successfully resolved. A bundle that is not resolved must neither export nor import packages. A bundle must have the necessary PackagePermission to participate in the sharing of a package.

A bundle declares the resources it offers to provide to other bundles using Export-Package manifest headers, and declares the resources it needs using Import-Package manifest headers.

2.4.3 Exporting Packages

The Export-Package manifest header allows a bundle to export Java packages to the OSGi environment, exposing the packages to other bundles.

The Framework must guarantee that classes and resources in the exported package's namespace are loaded from the exporting bundle. Additionally, the package's classes and resources must be shared among bundles that import the package, see *Importing Packages* on page 22.

If more than one bundle declares the same package in its Export-Package manifest header, the Framework controls the selection of the exporting bundle. The Framework must pick the bundle offering the highest version of the declared package.

In order to export a package, a bundle must have PackagePermission[EXPORT].

The Export-Package manifest header must conform to the following syntax:

```
Export-Package =
  package-description
  *( "," package-description )

package-description =
  package-name *( ";" parameter )

package-name =
  <fully qualified package name>

parameter =
  attribute "=" value

attribute = token
value     = token | quoted-string
```

The only package-description parameter recognized by the Framework is the attribute specification-version. Its string value must conform to the semantics described in the [9] *The Java 2 Package Versioning Specification.*

As an example, the following Export-Package manifest header declares that the bundle provides all classes defined by version 2.1 of the javax.servlet and javax.servlet.http packages.

```
Export-Package: javax.servlet;
    specification-version="2.1",
  javax.servlet.http;
    specification-version="2.1"
```

2.4.4   Importing Packages

The Import-Package manifest header allows a bundle to request access to packages that have been exported by other bundles in the OSGi environment.

The Framework must guarantee that while a bundle is resolved, the bundle is only exposed to one version of a package it has imported.

The fully qualified package name must be declared in the bundle's Import-Package manifest header for all packages a bundle needs, except for package names beginning with:

```
java.
```

In order to be allowed to import a package, a bundle must have PackagePermission[IMPORT] or PackagePermission[EXPORT].

PackagePermission on page 105 for more information.

The Import-Package manifest header must conform to the following syntax:

```
Import-Package =
  package-description
  *( "," package-description )

package-description =
  package-name *( ";" parameter )

package-name =
  <fully qualified package name>

parameter  = attribute "=" value
attribute  = token
value      = token | quoted-string
```

The only package-description parameter recognized by the Framework is the attribute specification-version. Its string value must conform to the semantics described in the [9] *The Java 2 Package Versioning Specification.*

As an example, the following Import-Package manifest header requires that the bundle be resolved against the javax.servlet package version 2.1 or above:

```
Import-Package: javax.servlet;
    specification-version="2.1"
```

2.4.5   Importing a Lower Version Than Exporting

Exporting a package does not imply that the exporting bundle will actually use the classes it offers for export. Multiple bundles can offer to export the same package; the Framework must select only one of those bundles as the exporter.

A bundle will implicitly import the same package name and version level as it exports, and therefore a separate Import-Package manifest header for this package is unnecessary. If the bundle can function using a lower specification version of the package than it exports, then the lower version can be specified in an Import-Package manifest header.

2.4.6   Code Executed Before Started

Packages exported from a bundle are exposed to other bundles as soon as the bundle has been resolved. This condition could mean that another bundle might call methods in an exported package *before* the bundle exporting the package is started.

2.4.7   Recommended Export Strategy

Although a bundle can export all of its classes to other bundles, this practice is discouraged except in the case of particularly stable library packages that will need updating only infrequently. The reason for this caution is that the Framework may not be able to promptly reclaim the space occupied by the exported classes if the bundle is updated or uninstalled.

Bundle designs that separate interfaces from their implementations are strongly preferred. The bundle developer should put the interfaces into a separate Java package to be exported, while keeping the implementation classes in a different package hidden from the outside world.

If the same interface has multiple implementations in multiple bundles, the bundle developer can package the interface package into all of these bundles; the Framework must select one, and only one, of the bundles to export the package, and the interface classes must be loaded from that bundle. Interfaces with the same package and class name should have exactly the same method signatures. Because a modification to an interface affects all of its callers, interfaces should be carefully designed and remain stable once deployed.

2.4.8 Bundle Classpath

Intra-bundle classpath dependencies are declared in the Bundle-Classpath manifest header. This declaration allows a bundle to declare its internal classpath using one or more JAR files that it contains.

The Bundle-Classpath manifest header is a list of comma-separated file names. A file name can be either dot (" . ") or the path of a JAR file contained in the bundle's JAR file. The dot represents the bundle's JAR file itself.

Classpath dependencies must be resolved as follows:

- If a Bundle-Classpath header is not declared, the default value of dot (.) is used, which specifies the bundle's JAR file.
- If a Bundle-Classpath manifest header is declared and dot (.) is not an element of the classpath, the bundle's JAR file must not be searched. In this case, only the JAR files specified within the bundle's JAR file must be searched.

The Bundle-Classpath manifest header must conform to the following syntax:

```
Bundle-Classpath =
  path *( "," path )

path         =
  <path name of nested JAR file with
    "/"-separated components>
```

For example, the following declaration in a bundle's manifest file would expose all classes and resources stored in the JAR file, but also all classes and resources defined in servlet.jar, to the bundle:

```
Bundle-Classpath: .,
  lib/servlet.jar
```

2.5 Loading Native Code Libraries

If a bundle has a Bundle-NativeCode manifest header, the bundle contains native code libraries that must be available for the bundle to execute. When a bundle makes a request to load a native code library, the findLibrary method of the caller's classloader must be called to return the file path name in which the Framework has made the requested native library available.

The bundle must have the required RuntimePermission[loadLibrary.< library name>] in order to load native code in the OSGi environment.

The Bundle-NativeCode manifest header must conform to the following syntax:

```
Bundle-NativeCode =
   nativecode-clause *( "," nativecode-clause )

nativecode-clause =
   nativepaths *( ";" env-parameter )

nativepaths =
   nativepath *( ";" nativepath )

nativepath  = </ separated path>

env-parameter = (
   processordef     |
   osnamedef        |
   osversiondef     |
   languagedef
)

processordef     = "processor"  "=" value
osnamedef        = "osname"      "=" value
osversiondef     = "osversion"  "=" value
languagedef      = "language"   "=" value

value      = token  | quoted-string
```

If a Bundle-NativeCode clause contains duplicate env-parameter entries, the corresponding values must be or'ed together.

If multiple native code libraries need to be installed on one platform, they must be specified in the same clause for that platform.

The following is an example of native code declaration in a bundle's manifest file:

```
Bundle-NativeCode: /lib/http.DLL ;
   /lib/zlib.dll ;
      osname       = Windows95 ;
      osname       = Windows98 ;
      osname       = WindowsNT ;
      osversion    = "5.1" ;
      processor    = x86 ;
      language     = en ;
      language     = se ,
   /lib/solaris/libhttp.so ;
      osname       = Solaris ;
      osname       = SunOS ;
      processor    = sparc,
   /lib/linux/libhttp.so ;
      osname       = Linux ;
      processor    = mips
```

2.5.1   Native Code Algorithm

1. Pick the clauses that match the processor and operating system of the Framework host computer.
   - The specified processor (attribute processor) and operating system-names (attribute osname) are compared to the Framework properties org.osgi.framework.processor and org.osgi.framework.os.name respectively. Case differences are ignored and aliases of the osname or processor attributes may be used.
   - If no clause matches both the required processor and operating system, the bundle installation fails.

2. Pick the clauses that best match the operating system version of the Framework host computer.
   - The specified operating system version (attribute osversion) is compared to the value specified in the Framework property: org.osgi.framework.os.version.
   - If there is only one clause with an exact match, it can be used as the best match and returned.
   - If there is more than one clause that matches the property, they must be picked to perform the next step.
   - Operating system versions are taken to be backward compatible. If there is no exact match in the clauses, clauses with operating system versions lower than the value specified in org.osgi.framework.os.version must be picked as the best

> match. If there is only one clause that has a compatible operating system version, it can be used as the best match.
> - Otherwise, all clauses with compatible operating system versions will go through the next step.
> - If no clause has a matching or compatible operating system version, the clause that does not have an operating system version specified must be picked as the best match.
> - If that is not possible, the bundle installation fails.

3. Pick the clause that best matches the language of the Framework host computer.
   - The specified language (clause attribute language) is matched against the system property: org.osgi.framework.language
   - If more than one clause remains at that point, then the Framework is free to pick among them randomly.
   - If no clauses have the exact match with the value of the org.osgi.framework.language property, the clause that does not have a language specified must be picked as the best match.
   - If that is not possible, the bundle installation fails.

2.6   Finding Classes and Resources

Framework implementations must follow the rules defined in this section regarding class and resource loading to create a predictable environment for bundle developers.

A bundle's classloader responds to requests by the bundle to load a resource or class. The bundle's classloader uses a delegation model. Upon a request to load a resource or class, the following classloaders are searched for the first occurrence of the class or resource, in the following order:

1. The system classloader.

2. The classloader of the bundle that exports the shared package to which the resource belongs (if the package is shared and the caller has the appropriate PackagePermission[IMPORT]).

3. The bundle's own classloader. The bundle is searched in the order specified in the Bundle-Classpath manifest header. See *Bundle Classpath* on page 24.

A class loaded from a bundle always belongs to that bundle's ProtectionDomain object.

2.6.1   Resources

In order to have access to a resource in a bundle, appropriate permissions are required. A bundle must always be given the necessary permissions by the Framework to access the resources contained in its JAR file (these permissions are Framework implementation dependent), as well as permission to access any resources in shared packages for which the bundle has PackagePermission[IMPORT].

When findResource is called on a bundle's classloader, the caller is checked for the appropriate permission to access the resource. If the caller does not have the necessary permission, the resource is not accessible and null must be returned. If the caller has the necessary permission, then a URL object to the resource must be returned. Once the URL object is returned, no further permission checks are performed when the contents of the resource are accessed. The URL object must use a protocol defined by the Framework implementation, and only the Framework implementation must be able to construct such URL objects of this protocol. The external form of this URL object must also be defined by the implementation.

A resource in a bundle may also be accessed by using the Bundle.getResource method. This method calls getResource on the bundle's classloader to perform the search. The caller of Bundle.getResource must have AdminPermission.

2.7   The Bundle Object

For each bundle installed in the OSGi environment, there is an associated
Bundle object. The Bundle object for a bundle can be used to manage the bundle's lifecycle. A management bundle is responsible for managing the lifecycle of other bundles.

2.7.1   Bundle Identifier

The bundle identifier is unique and persistent and has the following properties:

- The identifier is of type long.
- Once its value is assigned to a bundle, that value must not be reused for another bundle, even if the original bundle is reinstalled.
- Its value must not change as long as the bundle remains installed.
- Its value must not change when the bundle is updated.

The Bundle interface defines a getBundleId() method for returning a bundle's Identifier attribute.

2.7.2   Bundle Location

The bundle location is the location string of the bundle that was specified when the bundle was installed. The Bundle interface defines a getLocation() method for returning a bundle's location attribute.

A location string uniquely identifies a bundle.

2.7.3   Bundle State

A bundle may be in one of the following states:

- INSTALLED – The bundle has been successfully installed.
- RESOLVED – All Java classes and/or native code that the bundle needs are available. This state indicates that the bundle is either ready to be started or has stopped.
- STARTING – The bundle is being started, and the BundleActivator.start method has been called and has not yet returned.
- STOPPING – The bundle is being stopped, and the BundleActivator.stop method has been called and has not yet returned.
- ACTIVE – The bundle has successfully started and is running.
- UNINSTALLED – The bundle has been uninstalled. It cannot move into another state.

Figure 10   *State diagram Bundle*

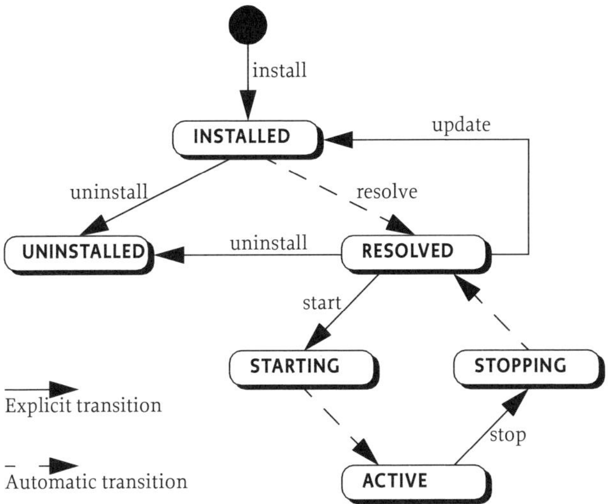

When a bundle is installed, it is stored in the persistent storage of the Framework and remains there until it is explicitly uninstalled. Whether a bundle has been started or stopped is recorded in the persistent storage of the Framework. A bundle that has been persistently recorded as started must be started whenever the Framework starts until the bundle is explicitly stopped.

The Bundle interface defines a getState() method for returning a bundle's state.

Bundle states are expressed as a bit-mask to conveniently determine the state of a bundle; a bundle can only be in one state physically at any time. The following code sample can be used to determine if a bundle is in the RESOLVED state:

```
if ((b.getState() & RESOLVED) != 0)
    ...
```

2.7.4 Installing Bundles

The BundleContext interface, which is given to the Bundle Activator of a bundle, defines the methods for installing a bundle:

- installBundle(String) – Installs a bundle from the specified location string.
- installBundle(String, InputStream) – Installs a bundle from the specified InputStream.

Every bundle is uniquely identified by its location string. If an installed bundle is using the specified location, the installBundle methods must return the Bundle object for that bundle.

The installation of a bundle in the Framework must be:

- *Persistent* – The bundle must remain installed across Framework and Java Virtual Machine invocations, until it is explicitly uninstalled.
- *Atomic* – The install method must completely install the bundle or, if the installation fails, the OSGi environment must be left in the same state as it was in before the method was called.

When installing a bundle, the Framework attempts to resolve the bundle's native code dependencies. If this attempt fails, the bundle must not be installed, see *Loading Native Code Libraries* on page 25.

Once a bundle has been installed, a Bundle object is created and all remaining lifecycle operations must be performed upon this object. The returned Bundle object can be used to start, stop, update, and uninstall the bundle.

2.7.5 Resolving Bundles

A bundle can enter the RESOLVED state when the Framework has successfully resolved the bundle's code dependencies. These dependencies include the bundle's:

- Classpath dependencies from its Bundle-Classpath manifest header.

- Native Code dependencies from its Bundle-NativeCode manifest header.
- Package dependencies from its Export-Package and Import-Package manifest headers.

If the bundle's dependencies are resolved, all the packages declared in the bundle's Export-Package manifest header are exported to the OSGi environment.

A bundle may be resolved at the Framework implementation's discretion once the bundle is installed.

2.7.6   Starting Bundles

The Bundle interface defines the start() method for starting a bundle. If this method succeeds, the bundle's state is set to ACTIVE and it remains in this state until it is stopped.

In order to be started, a bundle must be resolved. The Framework must attempt to resolve the bundle, if not already resolved, when trying to start the bundle. If the bundle fails to resolve, the start method must throw a BundleException.

If the bundle is resolved, the bundle must be activated by calling its Bundle Activator object, if any. The BundleActivator interface defines methods that the Framework invokes when it starts and stops the bundle.

To inform the OSGi environment of the fully qualified class name serving as its Bundle Activator, a bundle developer must declare a Bundle-Activator manifest header in the bundle's manifest file. The Framework must instantiate an object of this class and cast it to a BundleActivator. It must then call the start method to start the bundle.

The following is an example of a Bundle-Activator manifest header:

```
Bundle-Activator: com.acme.BA
```

A BundleActivator class must implement the BundleActivator interface and be declared public and have a public default constructor so an instance of it may be created with Class.newInstance.

Supplying a Bundle Activator is optional. For example, a library bundle that only exports a number of packages usually does not need to define a Bundle Activator. A bundle providing a service should do so, however, because this is the only way for the bundle to obtain its BundleContext object and get control when started.

The BundleActivator interface defines these methods for starting and stopping a bundle:

- `start(BundleContext)` – This method can allocate resources that a bundle needs and start threads, and also usually registers the bundle's services. If this method does not register any services, the bundle can register the services it needs at a later time, for example in a callback, as long as it is in the ACTIVE state.
- `stop(BundleContext)` – This method must undo all the actions of the `BundleActivator.start(BundleContext)` method.

2.7.7 Stopping Bundles

The `Bundle` interface defines the `stop()` method for stopping a bundle. This stops a bundle and sets the bundle's state to RESOLVED.

The `BundleActivator` interface defines a `stop(BundleContext)` method, which is invoked by the Framework to stop a bundle. This method must release any resources allocated since *activation*. All threads associated with the stopping bundle should be stopped immediately. The threaded code may no longer use Framework related objects (such as services and `BundleContext` objects) once its stop method returns.

This method may unregister services. However, if the stopped bundle had registered any services, either through its `BundleActivator.start` method, or while the bundle was in the ACTIVE state, the Framework must automatically unregister all registered services when the bundle is stopped.

The Framework guarantees that if a `BundleActivator.start` method has executed successfully, that same `BundleActivator` object must receive a call to its `BundleActivator.stop` method when the bundle is deactivated.

Packages exported by a bundle continue to be available to other bundles. This continued export implies that other bundles can execute code from a stopped bundle, and the designer of a bundle should assure that this is not harmful. Exporting only interfaces is one way to prevent this execution when the bundle is not started. Interfaces do not contain executable code so they cannot be executed.

2.7.8 Updating Bundles

The `Bundle` interface defines two methods for updating a bundle:

- `update()` – This method updates a bundle.
- `update(InputStream)` – This method updates a bundle from the specified InputStream.

The update process supports migration from one version of a bundle to a newer, backward-compatible version of the same bundle.

As an example, a bundle Newer, is backward compatible with another bundle, Older if:

- Newer provides at least the services provided by Older.
- Each service interface in Newer is compatible (as defined in [6] *The Java Language Specification*, Section 13.5) with its counterpart in Older.
- For any packages exported by Older, Newer must export the same package, which must be compatible with its counterpart in Older.

A Framework must guarantee that only one version of a bundle's classes is available at any time. If the updated bundle had exported any packages, those packages must not be updated; their old versions must remain exported until the org.osgi.service.admin.PackageAdmin.refreshPackages method has been called or the Framework is restarted.

2.7.9 Uninstalling Bundles

The Bundle interface defines a method for uninstalling a bundle from the Framework: uninstall(). This method causes the Framework to notify other bundles that the bundle is being uninstalled, and sets the bundle's state to
UNINSTALLED. The Framework must remove any resources related to the bundle that it is able to remove. This method must always uninstall the bundle from the persistent storage of the Framework.

Once this method returns, the state of the OSGi Environment must be the same as if the bundle had never been installed, unless the uninstalled bundle has exported any packages (via its Export-Package manifest header) and was selected by the Framework as the exporter of these packages.

If the bundle did export any packages, the Framework must continue to make these packages available to their importing bundles until one of the following conditions is satisfied:

- The org.osgi.service.admin.PackageAdmin.refreshPackages method has been called.
- The Framework is restarted.

2.8 The Bundle Context

The relationship between the Framework and its installed bundles is realized by the use of BundleContext objects. A BundleContext object represents the execution environment of a single bundle within the OSGi environment, and acts as a proxy to the underlying Framework.

A BundleContext object is created by the Framework when a bundle is started. The bundle can use this private BundleContext object for the following purposes:

- Installing new bundles into the OSGi environment. See *Installing Bundles* on page 30.
- Interrogating other bundles installed in the OSGi environment. See *Getting Bundle Information* on page 34.
- Obtaining a persistent storage area for the installed bundle. See *Persistent Storage* on page 35.
- Retrieving service objects of registered services. See *ServiceReference Objects* on page 38.
- Registering services in the Framework service. See *Registering Services* on page 39.
- Subscribing or unsubscribing to events broadcast by the Framework. See *Events* on page 50.

When a bundle is started, the Framework creates BundleContext object and provides this object as an argument to the start(BundleContext) method of the bundle's Bundle Activator. Each bundle is provided with its own BundleContext object; these objects should not be passed between bundles, as the BundleContext object is related to the security aspects of a bundle.

After the stop(BundleContext) method is called, the BundleContext object must no longer be used. Framework implementations must throw an exception if the BundleContext object is used after a bundle is stopped.

2.8.1   Getting Bundle Information

The BundleContext interface defines methods which can be used to retrieve information about bundles installed in the OSGi environment:

- getBundle() – Returns the single Bundle object associated with the BundleContext object.
- getBundles() – Returns an array of the bundles currently installed in the Framework.
- getBundle(long) – Returns the Bundle object specified by the unique identifier, or null if no matching bundle is found.

Bundle access is not restricted; any bundle can enumerate the set of installed bundles. Information that can identify a bundle, however (such as its location, or its header information), is only provided to callers that have AdminPermission.

2.8.2   Persistent Storage

The Framework should provide a private persistent storage area for each installed bundle on platforms with some file system support.

The BundleContext interface defines access to this storage in terms of the File class, which supports platform-independent definitions of file and directory names.

The BundleContext interface defines a method to access the private persistent storage area: getDataFile(String). This method takes a relative file name as an argument and translates it into an absolute file name in the bundle's persistent storage area and returns a File object. This method returns null if there is no support for persistent storage.

If a persistent area is given, the Framework must give the bundle the appropriate permissions to access this storage area.

2.8.3   Environment Properties

The BundleContext interface defines a method for returning information pertaining to Framework properties: getProperty(String). This method can be used to return the following Framework properties:

Property name	Description
org.osgi.framework.version	The version of the Framework.
org.osgi.framework.vendor	The vendor of the Framework implementation.
org.osgi.framework.language	The language being used is *ISO 639, International Standards Organization* See [10] *Codes for the Representation of Names of Languages* for valid values.
org.osgi.framework.processor	Processor name. The following table defines a list of processor names. New processors are defined on the OSGi web site. Names should be matched case insensitive.

Name	Aliases	Description
68k		68000 and up
ARM		Intel Strong ARM
Alpha		Compaq

Table 2   *Property Names*

Property name	**Description**		
	Ignite	psc1k	PTSC
	Mips		SGI
	PArisc		Hewlett Pack-ard
	PowerPC	power ppc	Motorola/IBM
	Sparc		SUN
	x86	pentium i386 i486 i586 i686	Intel

org.osgi.framework.os.name — The name of the operating system (OS) of the host computer. The following table defines a list of OS names. New OS names are defined on the OSGi web site. Names should be matched case insensitive.

Name	**Aliases**	**Description**
AIX		IBM
DigitalUnix		Compaq
FreeBSD		Free BSD
HPUX		Hewlett Pack-ard
IRIX		Sillicon Graph-ics
Linux		Open source
MacOS		Apple
Netware		Novell
OpenBSD		Open source
NetBSD		Open source
OS2	OS/2	IBM
QNX	procnto	QNX
Solaris		Sun Micro Sys-tems
SunOS		Sun Micro Sys-tems

Table 2   *Property Names*

Property name		Description	
	VxWorks		WindRiver Systems
	Windows95	Win95 Windows 95	Microsoft Windows 95
	Windows98	Win98 Windows 98	Microsoft Windows 98
	WindowsNT	WinNT Windows NT	Microsoft Windows NT
	WindowsCE	WinCE Windows CE	Microsoft Windows CE
	Windows2000	Win2000 Windows 2000	Microsoft Windows 2000

Table 2   *Property Names*

2.9 Services

In the OSGi environment, bundles are built around a set of cooperating services available from a shared service registry. Such an OSGi service is defined by its *service interface* and implemented as a *service object*.

Services are provided in the Framework as *service objects*. Service objects implement one or more service interfaces under which the service is registered with the Framework.

The semantics and behavior of the service object are defined by a service interface. This interface should be specified with as few implementation details as possible. OSGi has specified many interfaces for common needs and will specify more in the future.

The service object is owned by and runs within a bundle; this bundle must register the service with the Framework service registry so that the service's functionality is available to other bundles under control of the Framework.

Dependencies between the bundle owning the service and the bundles using it are managed by the Framework. For example, when a bundle is stopped, all the services registered with the Framework by that bundle must be automatically unregistered.

The Framework maps services to their underlying service objects, and provides a simple but powerful query mechanism that enables an installed bundle to request the services it needs. The Framework also provides an event mechanism so that bundles can receive events of service objects that are registered, modified, or unregistered.

2.9.1 ServiceReference Objects

In general, registered services are referenced through ServiceReference objects. This avoids creating unnecessary dynamic service dependencies between bundles.

A ServiceReference object can be stored and passed on to other bundles without the implications of dependencies. When a bundle wishes to use the service, it can be obtained by passing the ServiceReference object to BundleContext.getService(ServiceReference), see *Obtaining Services* on page 43.

A ServiceReference object encapsulates the properties and other meta information about the service object it represents. This meta information can be queried by a bundle to assist in the selection of a service that best suits its needs.

When a bundle queries the Framework service registry for services, if one or more of the registered services meet the request criteria, the Framework provides the requesting bundle with ServiceReference objects, rather than with the services themselves.

2.9.2 Service Interfaces

A *service interface* is the specification of the service's public methods.

In practice, a bundle developer creates a service object by implementing its service interface, and registers the service with the Framework service registry. Once a bundle has registered a service object under an interface/class name, the associated service can be acquired by bundles under that interface name, and its methods can be accessed by way of its service interface.

A bundle registers a service under the names of one or more interfaces or classes that the service object implements. Properties that describe the service object (vendor, version, etc.) can be specified as registration arguments. See *Properties* on page 40 for more information about these properties.

When requesting a service object from the Framework, a bundle can specify the name of the service interface that the requested service object must implement. In the request, the bundle may optionally specify a filter string to further narrow the search.

Many service interfaces are defined and specified by organizations such as OSGi. A service interface that has been accepted as a standard can be implemented and used by any number of bundle developers.

2.9.3 Registering Services

A bundle introduces a service into the service by registering its service object with the Framework service registry. A service object registered with the Framework is exposed to other bundles installed in the OSGi environment.

During registration, a service object can be given a set of key/value properties to support a sophisticated retrieval mechanism based on attribute comparisons.

Every registered service object has a unique ServiceRegistration object, and has one or more ServiceReference objects that refer to it. These ServiceReference objects expose the registration properties of the service object, including the set of service interfaces/classes it implements. The ServiceReference object can then be used to acquire a service object that implements the desired service.

The Framework permits bundles to register and unregister service objects dynamically. Therefore, a bundle is permitted to register service objects from the time its BundleActivator.start method is called until its BundleActivator.stop method is called and returns.

A bundle registers a service object with the Framework by calling one of the BundleContext.registerService methods on its BundleContext object:

- registerService(String, Object, Dictionary) – For a service object registered under a single service interface, of which it is an instance.
- registerService(String[], Object, Dictionary) – For a service object registered under multiple service interfaces, of which it is an instance.

The names of the service interface under which a bundle wants to register its service are provided as arguments to the BundleContext.registerService method. The Framework must ensure that the service object actually is an instance of all the service interfaces specified by the arguments.

The service object being registered may be further described by a Dictionary object, which contains the properties of the service as a collection of key/value pairs.

The service interface names under which a service object has been successfully registered are automatically added to the service object's properties under the key objectClass. This value must be set automatically by the Framework and any value provided by the bundle is overridden.

If the service object is successfully registered, the Framework returns a ServiceRegistration object to the caller. A service object can be unregistered only by the holder of its ServiceRegistration object (see unregister()). Every successful service object registration must yield a unique ServiceRegistration object, even if the same service object is registered multiple times.

Using the ServiceRegistration object is the only way to reliably change the service object's properties after it has been registered (see setProperties(Dictionary)). Modifying a service object's Dictionary object after the service obect is registered may not have any effect on the service properties.

2.9.4 Properties

Properties hold information as key/value pairs. The key is a String object and the value can be any type, but using a type recognized by Filter objects for comparison is recommended (see *Filters* on page 45 for a list). Multiple values for the same key are supported with arrays ([ ]) and Vector objects.

The values of properties should be limited to primitive or standard Java types to prevent unwanted inter-bundle dependencies. The Framework cannot detect dependencies that are created by the exchange of objects between bundles.

The key of a property is not case sensitive. ObjectClass, OBJECT-CLASS and objectclass all are the same property key. A Framework must, however, return the key in ServiceReference.getProper-tyKeys in exactly the same case as it was last set. When a Dictionary object is passed that contains keys that only differ in case, the Framework must raise an exception.

The properties of a ServiceRegistration object are intended to provide information *about* the service object. The properties should normally not be used to participate in the actual function of the service. Modifying the properties for the service registration is a potentially expensive operation; for example, a Framework may pre-process the properties into an index during registration, to speed up later look ups.

The Filter interface supports complex filtering and may be used to find matching service objects. Therefore, all properties share a single namespace in the Framework service registry. As a result, it is important to use descriptive names or formal definitions of shorter names to prevent conflicts. Several OSGi specifications reserve parts of this namespace. All properties starting with service. and the property objectClass are reserved for use by OSGi.

Table 3 Standard Framework Service Registry Properties contains a list of pre-defined properties. The first column contains the exact name of each property, and the next column contans a reference to the Constants interface where this property is defined in a constant with that name.

Property Key	Type	Constants	Property Description
objectClass	String[ ]	OBJECTCLASS	The objectClass property contains the set of classes and interfaces under which a service object is registered with the Framework. The Framework sets this property automatically. The Framework must guarantee that when a service object is retrieved with BundleContext.getService(ServiceReference) it can be cast to any of these classes or interfaces.
service.description	String	SERVICE_DESCRIPTION	The service.description property is intended to be used as documentation and is optional. The registering bundle can provide this property. Frameworks and bundles can use this property to provide a short description of a registered service object. The purpose is mainly for debugging because there is no support for localization.

Table 3	*Standard Framework Service Registry Properties*

Property Key	Type	Constants	Property Description
service.id	Long	SERVICE_ID	Every registered service object is assigned a service.id by the Framework, which is added to the service object's properties. The Framework assigns a unique value to every registered service object that is larger than values provided to all previously registered service objects.
service.pid	String	SERVICE_PID	The service.pid property optionally identifies a persistent, unique name for the service object. This name must be assigned by the bundle registering and should be a unique String object. Every time this service object is registered, including after a restart of the Framework, this service object should be registered under the same service.pid property. The value can be used by other bundles to persistently store information about this service object.
service.ranking	Integer	SERVICE_RANKING	When registering a service object, a bundle may optionally specify a service.ranking number as one of the service object's properties. If multiple qualifying service interfaces exist, a service's SERVICE_RANKING number, along with its SERVICE_ID, determine which service object is returned by the Framework.
service.vendor	String	SERVICE_VENDOR	This optional property can be used by the bundle registering the service object to define the vendor. The service.vendor property can be used by Framework to identify the service object.

Table 3   *Standard Framework Service Registry Properties*

2.9.5 Security Check

The process of registering a service object is subject to a security check. The registering bundle must have ServicePermission[REGIS-TER] to register the service object under all the service interfaces specified.

Otherwise, the service object must not be registered, and a SecurityException must be thrown. See *Permission Types* on page 54 for more information.

2.9.6 Obtaining Services

In order to use a service object and call its methods, a bundle must first obtain a ServiceReference object referencing the service object. The BundleContext interface defines two methods a bundle can call to obtain ServiceReference objects from the Framework:

- getServiceReference(String) – This method returns a Service-Reference object to a service object that implements, and was registered under, the name of the service interface specified as String. If multiple such service objects exist, the service object with the highest SERVICE_RANKING is returned. If there is a tie in ranking, the service object with the lowest SERVICE_ID (the service object that was registered first) is returned.
- getServiceReferences(String, String) – This method returns an array of ServiceReference objects that:
  - Implement and were registered under the service interface.
  - Satisfy the search filter specified. The filter syntax is further explained in *Filters* on page 45.

Both methods must return null if no matching service objects are returned. Otherwise the caller receives one or more ServiceReference objects. These objects can be used to retrieve properties of the underlying service object, or they can be used to obtain the actual service object via the BundleContext object.

2.9.7 Getting Service Properties

To allow for interrogation of service objects, the ServiceReference interface defines these two methods:

- getPropertyKeys() – Returns an array of the property keys that are available.
- getProperty(String) – Returns the value of a property.

Both of these methods must continue to provide information about the referenced service object, even after it has been unregistered from the Framework. This requirement can be useful when a ServiceReference object is stored with the Log Service.

2.9.8   **Getting Service Objects**

The BundleContext object is required to obtain the actual service object so that the Framework can account for the dependencies. If a bundle retrieves a service object, it becomes dependent upon the registered service object. This dependency is tracked by the Bundle-Context object used to obtain the service object, and is one reason that it is important to be careful when sharing BundleContext objects with other bundles.

The method BundleContext.getService(ServiceReference) returns an object that implements the interfaces as defined by the object-Class property.

This method has the following characteristics:

- Returns null if the underlying service object has been unregistered.
- Determines if the caller has ServicePermission[GET], to get the service object using at least one of the service interfaces under which the service was registered. This permission check is necessary so that ServiceReference objects can be passed around freely, without compromising security.
- Increments the bundle usage count of the service object by one.
- If the service object implements the ServiceFactory interface it is not returned, but used to customize the service object for the calling bundle. If the bundle usage of the service object is one, the ServiceFactory object is called to obtain a new service. Else, a cached copy of this customized object is returned. See *Service Factories* on page 46 for more information about ServiceFactory objects.

Both of the BundleContext.getServiceReference methods require that the caller have the required ServicePermission[GET] to get the service object for the specified service interface names. If the caller lacks the required permission, these methods return null.

Getting a ServiceReference object from a ServiceRegistration object does not require any permission.

A ServiceReference object is valid only as long as the service object it references has not been unregistered, but the properties remain available.

A service object can be used by casting it to one of the Java object types of which it is an instance, and calling the methods defined by that type.

2.9.9   Stale References

The Framework must manage the dependencies between bundles. This management is, however, restricted to Framework structures. Bundles must listen to events generated by the Framework to clean up and remove stale references.

A stale reference is a reference to a Java object that belongs to the classloader of a bundle that is stopped or is associated with a service object that is unregistered. Standard Java does not provide any means to clean up these stale references, and bundle developers must analyze their code carefully to ensure that stale references are deleted.

Stale references are potentially harmful because they hinder the Java garbage collector from harvesting the classes, and possible the instances, of stopped bundles. The following means are available to minimize stale references:

- Use the Service Tracker, a utility class which greatly simplifies tracking service objects. See *Service Tracker Specification* on page 137.
- Implement service objects using the `ServiceFactory` interface. The methods in the `ServiceFactory` interface simplify tracking bundles that use these service objects. See *Service Factories* on page 46.
- Use indirection in the service object implementations. Service objects handed out to other bundles should use an pointer to the actual service object implementation. When the service object becomes invalid, the pointer is set to `null`, effectively removing the reference to the actual service object.

2.9.10   Registered Services

The `Bundle` interface defines these two methods for returning information pertaining to installed bundles:

- `getRegisteredServices()` – Returns the service objects that the bundle has registered with the Framework.
- `getServicesInUse()` – Returns the service objects that the bundle is using.

2.10   Filters

The Framework provides a `Filter` interface, and uses a search filter syntax in the `getServiceReference` methods that is based on an RFC 1960-based search filter string. Filter objects can be created by calling `BundleContext.createFilter(String)` with the chosen filter string. For the syntax, see *Filter* on page 99.

A Filter object can be used numerous times to determine if the match argument, a ServiceReference or a Dictionary object, matches the filter string that was used to create the Filter object.

A filter matches a key that has multiple values if it matches at least one of those values. For example,

```
Dictionary dict = new Hashtable();
dict.put( "cn", new String[] { "a", "b", "c" } );
```

The dict will match true against a filter with "(cn=a)" but also "(cn=b)".

Comparison of values is dependent on the type of the value. String objects are compared differently than numbers, and it is possible for a key be associated with multiple values. Note that keys in the match argument must always be String objects. The comparison is defined by the object type of the key's value.

The Filter.toString method must always return the filter string in canonical form. Unnecessary white space must be removed, although that does not make filters completely canonical, because the ordering of sub-expressions is not defined.

2.11 Service Factories

A Service Factory allows customization of the service object that is returned when a bundle calls BundleContext.getService(Service-Reference).

Normally, the service object that is registered by a bundle is returned. If, however, the service object that is registered implements the ServiceFactory interface, the Framework must call methods on this object to create a unique service object for each bundle that gets the service.

When the service object is no longer used by a bundle – for example, when that bundle is stopped – then the Framework must notify the ServiceFactory object.

ServiceFactory objects help manage bundle dependencies that are not explicitly managed by the Framework. By binding a returned service object to the requesting bundle, the service object can listen to events related to that bundle and remove objects (for example, listeners) registered by that bundle when it is stopped. Normally, listening to events is not even necessary, because the Framework must inform the ServiceFactory object when a service object is released by a bundle, which happens automatically when a bundle is stopped.

The ServiceFactory interface defines these the following methods:

- getService(Bundle, ServiceRegistration) – Is called by the
  Framework if a call is made to BundleContext.getService and:
  - The specified ServiceReference argument points to a service
    object that implements the ServiceFactory interface.
  - The bundle's usage count of that service object is zero; that is,
    the bundle currently does not have any dependencies on the
    service object.

  The call to BundleContext.getService must be routed by the
  Framework to this method, passing to it the Bundle object of the
  caller. The Framework must cache the mapping of the requesting
  bundle-to-service, and return the cached service object to the
  bundle on future calls to BundleContext.getService, as long as
  the requesting bundle's usage count of the service object is
  greater than zero.

  The Framework must check the service object returned by this
  method; if it is not an instance of all the classes named when the
  service factory was registered, null is returned to the caller that
  called getService.
- ungetService(Bundle, ServiceRegistration, Object) – This
  method is called by the Framework if a call is made to Bundle-
  Context.ungetService and:
  - The specified ServiceReference argument points to a service
    object that implements the ServiceFactory interface.
  - The bundle's usage count for that service object must drop to
    zero after this call returns; that is, the bundle is about to
    release its last dependency on the service object.

  The call to BundleContext.ungetService must be routed by the
  Framework to this method, so the ServiceFactory object can
  release the service object previously created.

  Additionally, the cached copy of the previously created service
  object must be unreferenced by the Framework so it may be garbage collected.

2.12   Importing and Exporting Services

The Export-Service manifest header, explained in *Manifest Headers*
on page 18, declares the interfaces that a bundle may register. It provides advisory information that is not used by the Framework. This
header is intended for use by server-side management tools.

The Export-Service manifest header must conform to the following
syntax:

```
Export-Service =
  class-name * ( "," class-name )
```

```
class-name = <fully qualified class name>
```

The Import-Service manifest header declares the interfaces the bundle may use. It provides advisory information that is not used by the Framework. This header is also intended for use by server-side management tools.

The Import-Service manifest header must conform to the following syntax:

```
Import-Service =
  class-name *( "," class-name )

class-name = <fully qualified class name>
```

2.13 Releasing Services

In order for a bundle to release a service object, it must remove the dynamic dependency on the bundle that registered the service object. The Bundle
Context interface defines a method to release service objects: unget-Service(ServiceReference). A reference to the service object is passed as the argument of this method.

This method returns a boolean value:

- false if the bundle's usage count of the service object is already zero when the method is being called, or the service object has already been unregistered.
- true if the bundle's usage count of the service object was more than 0 before it was decremented.

2.14 Unregistering Services

The ServiceRegistration interface defines the unregister() method to unregisters the service object. This removes the ServiceRegistration object from the Framework service registry. The ServiceReference object for this ServiceRegistration object can no longer be used to access the service object.

The fact that this method is on the ServiceRegistration object ensures that only the bundle holding this object can unregister the associated service object. The bundle that unregisters a service object, however, might not be the same bundle that registered it. As an example, the registering bundle could have passed the Service-Registration object to another bundle, endowing that bundle with the responsibility of unregistering the service object. Passing ServiceRegistration objects should be done with caution.

After ServiceRegistration.unregister successfully completes, the service object is:

- Completely removed from the Framework service registry. As a consequence, ServiceReference objects obtained for that service object can no longer be used to access the service object. Calling BundleContext.getService method with the ServiceReference object must return null.
- Unregistered, even if other bundles had dependencies upon it. Bundles must be notified of the unregistration through the publishing of a ServiceEvent object of type ServiceEvent.UNREGIS-TERING. This event is sent synchronously in order to give bundles the opportunity to release the service object.
  After receiving an event of type ServiceEvent.UNREGISTERING the bundle should release the service object and releasing any references it has to this object, so that the service object can be garbage collected by the Java Virtual Machine.
  For each bundle whose usage count for the service object remains greater than zero after all invoked ServiceListener objects have returned, the usage count must be set to zero and the service object must be released.

2.15 Configurable Services

A Configurable service is one that can be configured dynamically at runtime, to change its behavior. As an example, a configurable Http Service may support an option to set the port number it should listen.

A service object is administered as configurable by implementing the
Configurable interface, which has one method: getConfiguration-Object(). This method returns an Object instance that holds the configuration data of the service. As an example, a configuration object could be implemented as a Java Bean.

The configuration object handles all the configuration aspects of a service so that the service object itself does not have to expose its configuration properties.

Before returning the configuration object, getConfigurationObject should check that the caller has the required permission to access and manipulate it, and if not, it should throw a SecurityException. Note that the required permission is implementation-dependent.

The Configurable interface is a minimalistic approach to configuration management. The OSGi also has a more comprehensive Configuration Management specification. The Configurable service is intended to be superseded by the Configuration Admin service, see *Configuration Admin Service Specification* on page 223.

2.16 Events

The OSGi Framework supports the following types of events:

- ServiceEvent – Reports registration, unregistration, and property changes for service objects. All events of this kind must be delivered synchronously.
- BundleEvent – Reports changes in the lifecycle of bundles.
- FrameworkEvent – Reports that the Framework is started, or has encountered errors.

2.16.1 Listeners

A listener interface is associated with each type of event.

- ServiceListener – Called with an event of type ServiceEvent when a service object has been registered or modified, or is in the process of unregistering.
- BundleListener and SynchronousBundleListener – Called with an event of type BundleEvent when a bundle has been installed, started, stopped, updated, or uninstalled. SynchronousBundleListener objects are called synchronously during the processing of the event, and must be called before the BundleListener objects are called.
- FrameworkListener – Called with an event of type FrameworkEvent when the Framework starts or when asynchronous errors occur.

A security check is performed for each registered listener when a ServiceEvent occurs. The listener must not be called unless it has the required permission to get the service object.

BundleContext interface methods are defined which can be used to add and remove each type of listener.

A bundle that uses a service object should register a ServiceListener object to track the availability of the service object, and take appropriate action when the service object is unregistering.

Events are asynchronously delivered, unless otherwise stated, meaning that they are not necessarily delivered by the same thread that generated the event. Which thread is used to call an event listener is not defined.

2.16.2 Delivering Events

When delivering an event asynchronously, the Framework:

- Collects a snapshot of the listener list at the time the event is published (rather than doing so in the future just prior to event delivery), so that listeners do not enter the list after the event happened, but before the event is delivered.
- Ensures that listeners on the list at the time the snapshot is taken still belong to active bundles at the time the event is delivered.

If the Framework did not capture the current listener list when the event was published, but instead waited until just prior to event delivery, then it would be possible for a bundle to have started and registered a listener, and the bundle could see its own BundleEvent.INSTALLED event, which would be an error.

The following three scenarios illustrate this concept.

1. Scenario 1 event sequence:
   - Event A is published.
   - Listener 1 is registered.
   - Asynchronous delivery of Event A is attempted
   Expected Behavior: Listener 1 must not receive Event A, because it was not registered at the time the event was published.

2. Scenario 2 event sequence:
   - Listener 2 is registered.
   - Event B is published.
   - Listener 2 is unregistered.
   - Asynchronous delivery of Event B is attempted.
   Expected Behavior: Listener 2 receives EventB, because Listener2 was registered at the time Event B was published.

3. Scenario 3 event sequence:
   - Listener 3 is registered.
   - Event C is published.
   - The bundle that registered Listener 3 is stopped.
   - Asynchronous delivery of Event C is attempted.
   Expected Behavior: Listener 3 must not receive Event C, because its Bundle Context object is invalid.

2.17 Security

The Framework security paradigm is based on the Java 2 specification. If security checks are performed, they must be done according to [8] *The Java Security Architecture for JDK 1.2*. It is assumed that the reader is familiar with this specification.

The Java platform on which the Framework runs must provide the Java Security APIs necessary for Java 2 permissions. On resource-constrained platforms, these Java Security APIs may be stubs that allow the bundle classes to be loaded and executed, but the stubs never actually perform the security checks.

Many of the Framework methods require the caller to explicitly have certain permissions when security is enabled. Services may also have permissions specific to them that provide more finely grained control over the operations that they are allowed to perform. Thus a bundle that exposes service objects to other bundles may also need to define permissions specific to the exposed service objects.

For example, the User Admin service has an associated UserAdmin-Permission class that is used to control access to this service.

2.17.1 Permission Checks

When a permission check is done, java.security.AccessController should check all the classes on the call stack to ensure that every one of them has the permission being checked.

Because service object methods often allow access to resources to which only the bundle providing the service object normally has access, a common programming pattern uses java.security.Access-Controller.doPrivileged in the implementation of a service object. The service object can assume that the caller is authorized to call the service object because a service can only be obtained with the appropriate ServicePermission[GET]. It should therefore use only its own permissions when it performs its function.

As an example, the dial method of a fictitious PPP Service accesses the serial port to dial a remote server and start up the PPP daemon. The bundle providing the PPP Service will have permission to execute programs and access the serial port, but the bundles using the PPP Service may not have those permissions.

When the dial method is called, the first check will be to ensure that the caller has permission to dial. This check is done with the following code:

```
SecurityManager sm = System.getSecurityManager();
```

```
if ( sm != null )
   sm.checkPermission( new
com.acme.ppp.DialPermission() );
```

If the permission check does not throw an exception, the dial method must now enter a privileged state to actually cause the modem to dial and start the PPP daemon as shown in the following example.

```
Process proc = (Process)
   AccessController.doPrivileged( new
PrivilegedAction()  {
      public Object run() {
         Process proc = null;
         if ( connectToServer() )
            proc = startDaemon();
         return proc;
      }
   }
);
```

For alternate ways of executing privileged code, see [8] *The Java Security Architecture for JDK 1.2*

2.17.2   Privileged Callbacks

The following interfaces define bundle callbacks that are invoked by the Framework:

- BundleActivator
- ServiceFactory
- Bundle-, Service-, and FrameworkListener.

When any of these callbacks are invoked by the Framework, the bundle that caused the callback may still be on the stack. For example, when one bundle installs and then starts another bundle, the installer bundle may be on the stack when the BundleActivator.start method of the installed bundle is called. Likewise, when a bundle registers a service object, it may be on the stack when the Framework calls back the serviceChanged method of all qualifying ServiceListener objects.

Whenever any of these bundle callbacks try to access a protected resource or operation, the access contol mechanism should consider not only the permissions of the bundle receiving the callback, but also those of the Framework and any other bundles on the stack. This means that in these callbacks, bundle programmers normally would use doPrivileged calls around any methods protected by a permission check (such as getting or registering service objects).

In order to reduce the number of doPrivileged calls by bundle programmers, the Framework must perform a doPrivileged around any bundle callbacks. The Framework should have java.security.AllPermission, therefore a bundle programmer can assume that it is not further restricted except for its own permissions.

Bundle programmers do not need to use doPrivileged calls in their implementations of any callbacks registered with and invoked by the Framework.

For any other callbacks that are registered with a service object and therefore get invoked by the service-providing bundle directly, doPrivileged calls must be used in the callback implementation if the bundle's own privileges are to be exercised. Otherwise, the callback must fail if the bundle that initiated the callback lacks the required permissions.

2.17.3 Permission Types

The following permission types are defined by the Framework:

- AdminPermission – Enables access to the administrative functions of the Framework.
- ServicePermission – Controls service object registration and access.
- PackagePermission – Controls importing and exporting packages.

2.17.4 AdminPermission

An AdminPermission has no parameters associated with it and is always named admin. AdminPermission is required by all sensitive functions. AdminPermission has no actions.

2.17.5 ServicePermission

A ServicePermission has the following parameters.

- *Interface Name* – The interface name may end with a wildcard to match multiple interface names. (See java.security.BasicPermission for a discussion of wildcards.)
- *Action* – Supported actions are: REGISTER – Indicates that the permission holder may register the service object, and GET – Indicates that the holder may get the service.

When an object is being registered as a service object using Bundle Context.registerService, the registering bundle must have the ServicePermission to register all of the named classes, see *Registering Services* on page 39.

When a ServiceReference object is obtained from the service registry using BundleContext.getServiceReference or BundleContext.getServiceReferences respectively, the calling bundle must have the required ServicePermission[GET] to get the service object with the named class (see *ServiceReference Objects* on page 38).

When a service object is obtained from a ServiceReference object using BundleContext.getService(ServiceReference), the calling code must have the required ServicePermission[GET] to get the service object for at least one of the classes under which it was registered.

ServicePermission must be used as a filter for the service events generated by the Framework as well as for the methods to enumerate services, including Bundle.getRegisteredServices and Bundle.getServicesInUse. The Framework must assure that a bundle must not be able to detect the presence of a service that it does not have permission to access.

2.17.6 PackagePermission

Bundles can only import and export packages for which they have the required permission actions. A PackagePermission is valid across all versions of a package.

A PackagePermission has two parameters:

- The package that may be exported. A wildcard may be used. The granularity of the permission is the package, not the class name.
- The action, either IMPORT or EXPORT. If a bundle has permission to export a package, the Framework must automatically grant it permission to import the package.

A PackagePermission with * and EXPORT as parameters would be able to import and export any package.

2.17.7 Bundle Permissions

The Bundle interface defines a method for returning information pertaining to a bundle's permissions: hasPermission(Object). This method returns true if the bundle has the specified permission, and false if it does not or if the object specified by the argument is not an instance of java.security.Permission.

The parameter type is Object so that the Framework can be implemented on Java platforms that do not support Java 2 based security.

2.18 Framework Startup and Shutdown

A Framework implementation must be started before any services can be provided. The details of how a Framework should be started is not defined in this specification, and may be different for different implementations. Some Framework implementations may provide command line options, and others may read startup information from a configuration file. In all cases, Framework implementations must perform all of the following actions in exact order.

2.18.1 Startup

When the Framework is started, the following actions must occur:

1. Event handling is enabled; events can now be delivered to listeners. Events are discussed in *Events* on page 50.

2. The system bundle enters the STARTING state. More information about the system bundle can be found in *The System Bundle* on page 16.

3. A bundle's state is persistently recorded in the OSGi environment. When the Framework is restarted, all installed bundles previously recorded as being started must be started as described in the Bundle.start method. Any exceptions that occur during startup are broadcast as a Framework event of type FrameworkEvent.ERROR. Bundles and their different states are discussed in *The Bundle Object* on page 28.

4. The system bundle enters the ACTIVE state.

5. A Framework event of type FrameworkEvent.STARTED is broadcast.

2.18.2 Shutdown

The Framework will also need to be shut down on occasion. Shutdown can also be initiated by stopping the system bundle, covered in *The System Bundle* on page 16. When the Framework is shut down, the following actions must occur:

1. The system bundle enters the STOPPING state

2. All ACTIVE bundles are suspended as described in the Bundle.stop method, except that their persistently recorded state indicates that they must be restarted when the Framework is next started. Any exceptions that occur during shutdown are broadcast as a Framework event of type FrameworkEvent.ERROR.

3. Event handling is disabled.

2.19   The Framework on JDK 1.1

The Framework specification was authored assuming a Java 2 based runtime environment. This section addresses issues in implementing and deploying the OSGi Framework on JDK 1.1 based runtime environments.

2.19.1   ClassLoader.getResource

In JDK 1.1, the ClassLoader class does not provide the findResource method. Therefore, references to the findResource method in this document should be taken to refer to the getResource method.

2.19.2   ClassLoader.findLibrary

Java 2 introduced the findLibrary method, which allows classloaders to participate in the loading of native libraries. In JDK 1.1, all native libraries must be available on a single, global library path. Therefore, native libraries from different bundles have to reside in the same directory. If libraries have the same name, unresolvable conflicts may occur.

2.19.3   Resource URL

A bundle's classloader returns resource URL objects which use a Framework implementation-specific URLStreamHandler subclass to capture security information about the caller.

Prior to Java 2 no constructor which took a URLStreamHandler object argument existed, requiring the Framework implementation to register a URLStreamHandler object. Conceivably, then, other code than the Framework implementation could create this type of URL objects with falsified security information, and is thus a security threat. Therefore, the Bundle.getResource() method cannot be implemented securely in Java versions prior to Java 2.

2.20   Changes since 1.0

2.20.1   System Bundle

This specification introduces the concept of a *system bundle*, which represents the Framework and serves the following purposes:

- It is responsible for registering *system services* like the Package Admin service, which are tightly integrated with the Framework.
- It appears as the exporter of any packages that are loaded off the system classpath (including org.osgi.framework).
- Its lifecycle methods handle the lifecycle of the framework.

The system bundle is described in *The System Bundle* on page 16.

2.20.2 Service Properties

New service properties: service.id, service.pid and service.ranking. Further information about these constants can be found in *Properties* on page 40.

2.20.3 New Bundle Manifest Header Attributes

Two new bundle manifest header attributes are introduced for cataloguing and archiving purposes: Bundle-Copyright and Bundle-Category. Consult the Constants interface, or see *Constants* on page 92, for a list of standard manifest headers that may be declared in a bundle's manifest file.

2.20.4 New Framework Methods

- Bundle.getResource – Calls this bundle's classloader to search for the named resource. See *getResource(String)* on page 68.
- BundleContext.createFilter – Creates a Filter object from the given filter string, which may be used to match a ServiceReference or Dictionary object. See *createFilter(String)* on page 79.
- ServiceReference.getUsingBundles – Returns the bundles that are using the service referenced by this ServiceReference. See *ServiceReference* on page 113.

2.20.5 New Framework Classes

- Constants – Interface containing standard names for OSGi environment properties, service properties, and manifest headers. See *Constants* on page 92.
- Filter – Class which encapsulates a filter string and can be used to match a ServiceReference or Dictionary object. See *Filter* on page 99 and *Filters* on page 45.
- SynchronousBundleListener – A BundleListener interface which is notified of bundle lifecycle events in a synchronous fashion. See *Delivering Events* on page 51 and *SynchronousBundleListener* on page 117.

2.20.6 Bundle Classloader Delegation Model

This specification defines a standard delegation model for bundle classloaders, in order to make bundle classloading and bundle resource access consistent across Framework implementations from different vendors.

A bundle's classloader first delegates the request to load a class or access a resource to the system classloader, before searching any shared packages and the bundle itself (in this order) for the requested class or resource. Any packages on the system classpath appear to be exported by the system bundle. For further information see *Finding Classes and Resources* on page 27.

2.20.7 Replacing and Removing Exported Packages

In the previous release, an exported package would either remain exported until the Framework was shut down, or it could be replaced (during a bundle update) or removed (during a bundle uninstall) while the Framework was running, requiring the Framework to stop and resolve again all dependent bundles. The latter behavior was also known as "eager bundle update" and "eager bundle uninstall," respectively.

This specification eliminates the possibility of an eager bundle update and uninstall, meaning that when updating a bundle, its exported packages must not be updated. Likewise, when uninstalling a bundle, its exported packages must not be removed.

In order to replace or remove exported packages of bundles that were respectively updated or uninstalled, the refreshPackages method of the new org.osgi.service.packageadmin.PackageAdmin service must be called, which causes all dependent bundles to be resolved again. See *Package Admin Service Specification* on page 119 for further information.

2.20.8 Bundle URL objects

This specification introduces a new type of URL object to access resources in bundles. The URL object uses a protocol defined by the Framework implementation, and only the Framework implementation can construct URL objects of this protocol. This type of URL object is returned by the bundle classloader's findResource and the newly added org.osgi.framework.Bundle.getResource methods.

Before returning the URL object of a bundle resource, the caller is checked for the appropriate permission to access the resource (in the case of Bundle.getResource, the caller must have AdminPermission). Once the URL object is returned, no further permission checks are performed when the contents of the resource are accessed.

2.20.9 Optional Bundle Contents

This specification allows a bundle to carry optional resources (such as source code and documentation) in the OSGI-OPT directory in the bundle's JAR file.

See *Bundles* on page 16 for more information.

2.20.10 Clarifications

- Adding Listeners BundleContext.addBundleListener and BundleContext.addFrameworkListener now state that they do nothing and simply return if the BundleContext object of the bundle already contains the given listener.

- The version of BundleContext.addServiceListener that takes a filter string now states that if the context bundle already contains the given listener, that listener's filter (which may be null) is replaced with the given filter (which may be null).
A compatibility issue arises from this clarification: With the previous release, it was valid to register the same ServiceListener object with two mutually exclusive filters, e.g., "(objectClass=foo)" and "(objectClass=bar)". The listener would then be called in either case. In this specification, the Framework must only called the listener when the second filter matches.
- The use of spaces in the filter expression is clarified. See Filter.

2.20.11 Case Considerations of Service Properties

ServiceReference.getPropertyKeys has been clarified to say that it is case-preserving: that is, every key returned must be in exactly the same form as it was last set. Note that this change does not affect filter matching or ServiceReference.getProperty, which continue to treat property keys as case insensitive.

ServiceRegistration.setProperties and BundleContext.registerService now throw an IllegalArgumentException if the Dictionary object passed to them contains case variants of the same key name (the same is true for Filter.match, which was added in this release).

See *Properties* on page 40 for a further discussion of service properties.

2.20.12 Manifest Header Syntax

When specifying the syntax of various OSGi-specific manifest headers, SG 1.0 did not clearly distinguish between *tokens* and *strings*, which created ambiguities regarding spaces.

This specification uses an augmented Backus-Naur Form (BNF) similar to that used by [12] *RFC 822 Standard for the Format of ARPA Internet Text Messages* as the grammar for specifying OSGi-specific manifest headers. This is the same augmented BNF that is used in the specification of the HTTP 1.1 protocol, [4] *The Hypertext Transfer Protocol - HTTP/1.1*.

Finally, a Framework must not throw an exception when encountering any unknown manifest header attributes. Instead, it must make all headers in the main section accessible to bundle code by including them in the Dictionary object returned by Bundle.getHeaders.

For more information about the manifest syntax, see *Manifest Header Grammar* on page 18.

2.20.13 Event Delivery

The previous version specified that events of type ServiceEvent must be delivered synchronously, whereas all other types of events may be delivered asynchronously. Text has been added to the specification to describe in more detail the ramifications of delivering an event asynchronously; see *Delivering Events* on page 51.

2.20.14 Bundle Location

The bundle's location identifier must not change in a bundle's lifetime, not even during a bundle update. Therefore, the following section has been removed from org.osgi.framework.Bundle.getLocation:

The location identifier of the bundle may change during bundle update. Calling this method while the Framework is updating the bundle results in undefined behavior.

2.20.15 Native Code Selection Algorithm

The following sentence was removed from step 1 of the native code selection algorithm in SG 1.0:

If only one clause matches, it can be used, otherwise, remaining steps are executed.

The native code selection algorithm in SG 1.0 has been clarified to say that if multiple native libraries need to be installed on the same platform, they must be specified in the same clause for that platform, rather than in separate clauses. See *Loading Native Code Libraries* on page 25.

2.20.16 Registering Service Under a Single Interface

The BundleContext.registerService that registers a service under a single class name (instead of an array of class names) now reinforces that the value of the service's objectClass property will still be an array of strings, rather than just a String object. See *registerService(String, Object, Dictionary)* on page 87.

2.20.17 Bundle Callbacks Must Be Executed as Privileged Operations

Framework implementations must execute bundle callbacks as privileged operations: that is, they must perform a doPrivileged method around bundle callbacks. See *Privileged Callbacks* on page 53.

2.20.18   Bundle.uninstall

The publishing of the BundleEvent object has been moved to after
the bundle's state has been changed to UNINSTALLED. The reason
for this change is to make Bundle.uninstall consistent with the other
lifecycle methods in this regard.

2.21   org.osgi.framework

The OSGi Framework Package. Specification Version 1.1.

Bundles wishing to use this package must list the package in the
Import-Package header of the bundle's manifest. For example:

```
Import-Package: org.osgi.framework;specification-
version=1.1
```

Class Summary

Interfaces

Bundle	An installed bundle in the Framework.
BundleActivator	Customizes the starting and stopping of this bundle.
BundleContext	A bundle's execution context within the Framework.
BundleListener	A BundleEvent listener.
Configurable	Supports a configuration object.
Constants	Defines standard names for the OSGi environment property, service property, and Manifest header attribute keys.
Filter	An RFC 1960-based Filter.
FrameworkListener	A FrameworkEvent listener.
ServiceFactory	Allows services to provide customized service objects in the OSGi environment.
ServiceListener	A ServiceEvent listener.
ServiceReference	A reference to a service.
ServiceRegistration	A registered service.
SynchronousBundleListener	A synchronous BundleEvent listener.

Classes

AdminPermission	Indicates the caller's authority to perform lifecycle operations on or to get sensitive information about a bundle.

Class Summary

BundleEvent	A Framework event describing a bundle lifecycle change.
FrameworkEvent	A general Framework event.
PackagePermission	A bundle's authority to import or export a package.
ServiceEvent	A service lifecycle change event.
ServicePermission	Indicates a bundle's authority to register or get a service.

Exceptions

BundleException	A Framework exception used to indicate that a bundle life-cycle problem occurred.
InvalidSyntaxException	A Framework exception.

2.21.1 AdminPermission

**public final class AdminPermission extends
java.security.BasicPermission**

*All Implement-
ed Interfaces:*   java.security.Guard, java.io.Serializable

Indicates the caller's authority to perform lifecycle operations on or
to get sensitive information about a bundle.

AdminPermission has no actions or target.

The hashCode() method of AdminPermission is inherited from
java.security.BasicPermission. The hash code it returns is the hash
code of the name "AdminPermission", which is always the same for
all instances of AdminPermission.

2.21.1.1 Constructors

public **AdminPermission**()

Creates a new AdminPermission object with its name set to "Admin-
Permission".

public **AdminPermission**(java.lang.String name, java.lang.String
actions)

Creates a new AdminPermission object for use by the Policy object to
instantiate new Permission objects.

Parameters: name - Ignored; always set to "AdminPermission".

actions - Ignored.

2.21.1.2   **Methods**

public boolean **equals**(java.lang.Object obj)

Determines the equality of two Admin Permission objects.

Two Admin Permission objects are always equal.

Overrides: java.security.BasicPermission.equals(java.lang.Object) in class
java.security.BasicPermission

Parameters: obj - The object being compared for equality with this object.

Returns: true if obj is an Admin Permission; false otherwise.

public boolean **implies**(java.security.Permission p)

Determines if the specified permission is implied by this object.

This method returns true if the specified permission is an instance
of Admin Permission.

Overrides: java.security.BasicPermission.implies(java.security.Permission) in
class java.security.BasicPermission

Parameters: p - The permission to interrogate.

Returns: true if the permission is an instance of this class; false otherwise.

public java.security.PermissionCollection
   newPermissionCollection()

Returns a new Permission Collection object suitable for storing
Admin Permissions.

Overrides: java.security.BasicPermission.newPermissionCollection() in class
java.security.BasicPermission

Returns: A new Permission Collection object.

2.21.2   **Bundle**

public interface Bundle

An installed bundle in the Framework.

A Bundle object is the access point to define the life cycle of an
installed bundle. Each bundle installed in the OSGi environment
will have an associated Bundle object.

A bundle will have a unique identity, a long, chosen by the Framework. This identity will not change during the life cycle of a bundle,
even when the bundle is updated. Uninstalling and then reinstalling
the bundle will create a new unique identity.

A bundle can be in one of six states:

- UNINSTALLED
- INSTALLED
- RESOLVED
- STARTING
- STOPPING
- ACTIVE

Values assigned to these states have no specified ordering; they represent bit values that may be ORed together to determine if a bundle is in one of the valid states.

A bundle should only execute code when its state is one of STARTING, ACTIVE, or STOPPING. An UNINSTALLED bundle can not be set to another state; it is a zombie and can only be reached because invalid references are kept somewhere.

The Framework is the only entity that is allowed to create Bundle objects, and these objects are only valid within the Framework that created them.

2.21.2.1   **Fields**

public static final int **ACTIVE**

This bundle is now running.

A bundle is in the ACTIVE state when it has been successfully started.

The value of ACTIVE is 0x00000020.

public static final int **INSTALLED**

This bundle is installed but not yet resolved.

A bundle is in the INSTALLED state when it has been installed in the Framework but cannot run.

This state is visible if the bundle's code dependencies are not resolved. The Framework may attempt to resolve an INSTALLED bundle's code dependencies and move the bundle to the RESOLVED state.

The value of INSTALLED is 0x00000002.

public static final int **RESOLVED**

This bundle is resolved and is able to be started.

A bundle is in the RESOLVED state when the Framework has successfully resolved the bundle's dependencies. These dependencies include:

- The bundle's class path from its BUNDLE_CLASSPATH Manifest header.
- The bundle's native language code from its BUNDLE_NATIVECODE Manifest header.
- The bundle's package dependencies from its EXPORT_PACKAGE and IMPORT_PACKAGE Manifest headers.

Note that the bundle is not active yet. A bundle must be put in the RESOLVED state before it can be started. The Framework may attempt to resolve a bundle at any time.

The value of RESOLVED is 0x00000004.

public static final int **STARTING**

This bundle is in the process of starting.

A bundle is in the STARTING state when the start() method is active. A bundle will be in this state when the bundle's start(BundleContext) is called. If this method completes without exception, then the bundle has successfully started and will move to the ACTIVE state.

The value of STARTING is 0x00000008.

public static final int **STOPPING**

This bundle is in the process of stopping.

A bundle is in the STOPPING state when the stop() method is active. A bundle will be in this state when the bundle's stop(BundleContext) method is called. When this method completes the bundle is stopped and will move to the RESOLVED state.

The value of STOPPING is 0x00000010.

public static final int **UNINSTALLED**

This bundle is uninstalled and may not be used.

The UNINSTALLED state is only visible after a bundle is uninstalled; the bundle is in an unusable state and all references to the Bundle object should be released immediately.

The value of UNINSTALLED is 0x00000001.

2.21.2.2   **Methods**

public long **getBundleId**()

Returns this bundle's identifier. The bundle is assigned a unique identifier by the Framework when it is installed in the OSGi environment.

A bundle's unique identifier has the following attributes:

- Is unique and persistent.
- Is a long.
- Its value is not reused for another bundle, even after the bundle is uninstalled.
- Does not change while the bundle remains installed.
- Does not change when the bundle is updated.

This method will continue to return this bundle's unique identifier while this bundle is in the UNINSTALLED state.

Returns: The unique identifier of this bundle.

```
public java.util.Dictionary getHeaders()
    throws java.lang.SecurityException
```

Returns this bundle's Manifest headers and values. This method returns all the Manifest headers and values from the main section of the bundle's Manifest file; that is, all lines prior to the first blank line.

Manifest header names are case-insensitive. The methods of the returned Dictionary object will operate on header names in a case-insensitive manner.

For example, the following Manifest headers and values are included if they are present in the Manifest file:

```
Bundle-Name
Bundle-Vendor
Bundle-Version
Bundle-Description
Bundle-DocURL
Bundle-ContactAddress
```

This method will continue to return Manifest header information while this bundle is in the UNINSTALLED state.

Returns: A Dictionary object containing this bundle's Manifest headers and values.

Throws: java.lang.SecurityException - If the caller does not have the Admin-Permission, and the Java Runtime Environment supports permissions.

```
public java.lang.String getLocation()
    throws java.lang.SecurityException
```

Returns this bundle's location identifier.

The bundle location identifier is the location passed to installBundle(String) when a bundle is installed.

This method will continue to return this bundle's location identifier while this bundle is in the UNINSTALLED state.

Returns: The string representation of this bundle's location identifier.

Throws: java.lang.SecurityException - If the caller does not have the appropriate AdminPermission, and the Java Runtime Environment supports permissions.

```
public ServiceReference[] getRegisteredServices()
    throws java.lang.IllegalStateException
```

Returns this bundle's ServiceReference list for all services it has registered or null if this bundle has no registered services.

If the Java runtime supports permissions, a ServiceReference object to a service is included in the returned list only if the caller has the ServicePermission to get the service using at least one of the named classes the service was registered under.

The list is valid at the time of the call to this method, however, as the Framework is a very dynamic environment, services can be modified or unregistered at anytime.

Returns: An array of ServiceReference objects or null.

Throws: java.lang.IllegalStateException - If this bundle has been uninstalled.

See Also: ServiceRegistration, ServiceReference, ServicePermission

```
public java.net.URL getResource(java.lang.String name)
    throws java.lang.IllegalStateException
```

Find the specified resource in this bundle. This bundle's class loader is called to search for the named resource. If this bundle's state is INSTALLED, then only this bundle will be searched for the specified resource. Imported packages cannot be searched when a bundle has not been resolved.

Parameters: name - The name of the resource. See java.lang.ClassLoader.getResource for a description of the format of a resource name.

Returns: a URL to the named resource, or null if the resource could not be found or if the caller does not have the AdminPermission, and the Java Runtime Environment supports permissions.

Throws: java.lang.IllegalStateException - If this bundle has been uninstalled.

Since: 1.1

public ServiceReference[] **getServicesInUse**()
    throws java.lang.IllegalStateException

Returns this bundle's ServiceReference list for all services it is using or returns null if this bundle is not using any services. A bundle is considered to be using a service if its use count for that service is greater than zero.

If the Java Runtime Environment supports permissions, a Service-Reference object to a service is included in the returned list only if the caller has the ServicePermission to get the service using at least one of the named classes the service was registered under.

The list is valid at the time of the call to this method, however, as the Framework is a very dynamic environment, services can be modified or unregistered at anytime.

Returns: An array of ServiceReferences or null.

Throws: java.lang.IllegalStateException - If this bundle has been uninstalled.

See Also: ServiceReference, ServicePermission

public int **getState**()

Returns this bundle's current state.

A bundle can be in only one state at any time.

Returns: An element of UNINSTALLED, INSTALLED, RESOLVED, STARTING, STOPPING, ACTIVE.

public boolean **hasPermission**(java.lang.Object permission)
    throws java.lang.IllegalStateException

Determines if this bundle has the specified permissions.

If the Java Runtime Environment does not support permissions, this method always returns true.

permission is of type Object to avoid referencing the java.security.Permission class directly. This is to allow the Framework to be implemented in Java environments which do not support permissions.

If the Java Runtime Environment does support permissions, this bundle and all its resources including nested JAR files, belong to the same java.security.ProtectionDomain; that is, they will share the same set of permissions.

Parameters:

permission - The permission to verify.

Returns:

true if this bundle has the specified permission or the permissions possessed by this bundle imply the specified permission; false if this bundle does not have the specified permission or permission is not an instanceof java.security.Permission.

Throws:

java.lang.IllegalStateException - If this bundle has been uninstalled.

public void **start**()
    throws BundleException, java.lang.IllegalStateException,
    java.lang.SecurityException

Starts this bundle.

The following steps are required to start a bundle:

1. If this bundle's state is UNINSTALLED then an IllegalState-Exception is thrown.

2. If this bundle's state is STARTING or STOPPING then this method will wait for this bundle to change state before continuing. If this does not occur in a reasonable time, a BundleException is thrown to indicate this bundle was unable to be started.

3. If this bundle's state is ACTIVE then this method returns immediately.

4. If this bundle's state is not RESOLVED, an attempt is made to resolve this bundle's package dependencies. If the Framework cannot resolve this bundle, a BundleException is thrown.

5. This bundle's state is set to STARTING.

6. The start(BundleContext) method of this bundle's Bundle-Activator, if one is specified, is called. If the BundleActivator is invalid or throws an exception, this bundle's state is set back to RESOLVED.
   Any services registered by the bundle will be unregistered.
   Any services used by the bundle will be released.
   Any listeners registered by the bundle will be removed.
   A BundleException is then thrown.

7. If this bundle's state is UNINSTALLED, because the bundle was uninstalled while the BundleActivator.start method was running, a BundleException is thrown.

8. Since it is recorded that this bundle has been started, when the Framework is restarted this bundle will be automatically started.

9. This bundle's state is set to ACTIVE.

10.A bundle event of type STARTED is broadcast.

Preconditions

- getState() in {INSTALLED}, {RESOLVED}.

Postconditions, no exceptions thrown

- getState() in {ACTIVE}.
- BundleActivator.start() has been called and did not throw an exception.

Postconditions, when an exception is thrown

- getState() not in {STARTING}, {ACTIVE}.

Throws: BundleException - If this bundle couldn't be started. This could be because a code dependency could not be resolved or the specified BundleActivator could not be loaded or threw an exception.

java.lang.IllegalStateException - If this bundle has been uninstalled or this bundle tries to change its own state.

java.lang.SecurityException - If the caller does not have the appropriate AdminPermisson, and the Java Runtime Environment supports permissions.

```
public void stop()
    throws BundleException, java.lang.IllegalStateException,
    java.lang.SecurityException
```

Stops this bundle.

The following steps are required to stop a bundle:

1. If this bundle's state is UNINSTALLED then an IllegalState-Exception is thrown.

2. If this bundle's state is STARTING or STOPPING then this method will wait for this bundle to change state before continuing. If this does not occur in a reasonable time, a BundleException is thrown to indicate this bundle was unable to be stopped.

3. If this bundle's state is not ACTIVE then this method returns immediately.

4. This bundle's state is set to STOPPING.

5. Since it is recorded that this bundle has been stopped, Framework is restarted this bundle will not be automatically started.

6. The stop(BundleContext) method of this bundle's Bundle-Activator, if one is specified, is called. If this method throws an exception, it will continue to stop this bundle. A Bundle-

Exception will be thrown after completion of the remaining steps.

7. Any services registered by this bundle will be unregistered.

8. Any services used by this bundle will be released.

9. Any listeners registered by this bundle will be removed.

10.If this bundle's state is UNINSTALLED, because the bundle was uninstalled while the BundleActivator.stop method was running, a BundleException is thrown.

11.This bundle's state is set to RESOLVED.

12.A bundle event of type STOPPED is broadcast.

Preconditions

- getState() in {ACTIVE}.

Postconditions, no exceptions thrown

- getState() not in {ACTIVE, STOPPING}.
- BundleActivator.stop has been called and did not throw an exception.

Postconditions, when an exception is thrown

- None.

Throws: BundleException - If this bundle's BundleActivator could not be loaded or threw an exception.

java.lang.IllegalStateException - If this bundle has been uninstalled or this bundle tries to change its own state.

java.lang.SecurityException - If the caller does not have the appropriate AdminPermission, and the Java Runtime Environment supports permissions.

public void **uninstall**()
    throws BundleException, java.lang.IllegalStateException,
    java.lang.SecurityException

Uninstalls this bundle.

This method causes the Framework to notify other bundles that this bundle is being uninstalled, and then puts this bundle into the UNINSTALLED state. The Framework will remove any resources related to this bundle that it is able to remove.

If this bundle has exported any packages, the Framework will continue to make these packages available to their importing bundles until the PackageAdmin.refreshPackages method has been called or the Framework is relaunched.

The following steps are required to uninstall a bundle:

1. If this bundle's state is UNINSTALLED then an IllegalState-Exception is thrown.

2. If this bundle's state is ACTIVE, STARTING or STOPPING, this bundle is stopped as described in the Bundle.stop method. If Bundle.stop throws an exception, a Framework event of type ERROR is broadcast containing the exception.

3. This bundle's state is set to UNINSTALLED.

4. A bundle event of type UNINSTALLED is broadcast.

5. This bundle and any persistent storage area provided for this bundle by the Framework are removed.

Preconditions

- getState() not in {UNINSTALLED}.

Postconditions, no exceptions thrown

- getState() in {UNINSTALLED}.
- This bundle has been uninstalled.

Postconditions, when an exception is thrown

- getState() not in {UNINSTALLED}.
- This Bundle has not been uninstalled.

Throws: BundleException - If the uninstall failed.

java.lang.IllegalStateException - If this bundle has been uninstalled or this bundle tries to change its own state.

java.lang.SecurityException - If the caller does not have the appropriate AdminPermission, and the Java Runtime Environment supports permissions.

See Also: stop()

public void **update**()
    throws BundleException, java.lang.IllegalStateException,
    java.lang.SecurityException

Updates this bundle.

If this bundle's state is ACTIVE, it will be stopped before the update and started after the update successfully completes.

If the bundle being updated has exported any packages, these packages will not be updated. Instead, the previous package version will remain exported until the PackageAdmin.refreshPackages method has been has been called or the Framework is relaunched.

The following steps are required to update a bundle:

1. If this bundle's state is UNINSTALLED then an IllegalState-Exception is thrown.

2. If this bundle's state is ACTIVE, STARTING or STOPPING, the bundle is stopped as described in the Bundle.stop method. If Bundle.stop throws an exception, the exception is rethrown terminating the update.

3. The download location of the new version of this bundle is determined from either the bundle's BUNDLE_UPDATELOCATION Manifest header (if available) or the bundle's original location.

4. The location is interpreted in an implementation dependent manner, typically as a URL, and the new version of this bundle is obtained from this location.

5. The new version of this bundle is installed. If the Framework is unable to install the new version of this bundle, the original version of this bundle will be restored and a BundleException will be thrown after completion of the remaining steps.

6. This bundle's state is set to INSTALLED.

7. If this bundle has not declared an Import-Package header in its Manifest file (specifically, this bundle does not depend on any packages from other bundles), this bundle's state may be set to RESOLVED.

8. If the new version of this bundle was successfully installed, a bundle event of type UPDATED is broadcast.

9. If this bundle's state was originally ACTIVE, the updated bundle is started as described in the Bundle.start method. If Bundle.start throws an exception, a Framework event of type ERROR is broadcast containing the exception.

Preconditions

- getState() not in {UNINSTALLED}.

Postconditions, no exceptions thrown

- getState() in {INSTALLED, RESOLVED, ACTIVE}.
- This bundle has been updated.

Postconditions, when an exception is thrown

- getState() in {INSTALLED, RESOLVED, ACTIVE}.
- Original bundle is still used; no update occurred.

Throws: BundleException - If the update fails.

java.lang.IllegalStateException - If this bundle has been uninstalled or this bundle tries to change its own state.

java.lang.SecurityException - If the caller does not have the appropriate AdminPermission, and the Java Runtime Environment supports permissions.

See Also: stop(), start()

public void **update**(java.io.InputStream in)

Updates this bundle from an InputStream.

This method performs all the steps listed in Bundle.update(), except the bundle will be read from the supplied InputStream, rather than a URL.

This method will always close the InputStream when it is done, even if an exception is thrown.

Parameters: in - The InputStream from which to read the new bundle.

Throws: BundleException

See Also: update()

2.21.3   BundleActivator

public interface BundleActivator

Customizes the starting and stopping of this bundle.

BundleActivator is an interface that may be implemented when this bundle is started or stopped. The Framework can create instances of this bundle's BundleActivator as required. If an instance's Bundle-Activator.start method executes successfully, it is guaranteed that the same instance's BundleActivator.stop method will be called when this bundle is to be stopped.

BundleActivator is specified through the Bundle-Activator Manifest header. A bundle can only specify a single BundleActivator in the Manifest file. The form of the Manifest header is:

    Bundle-Activator: *class-name*

where class-name is a fully qualified Java classname.

The specified BundleActivator class must have a public constructor that takes no parameters so that a BundleActivator object can be created by Class.newInstance().

2.21.3.1   **Methods**

public void **start**(BundleContext context)
    throws java.lang.Exception

Called when this bundle is started so the Framework can perform the bundle-specific activities necessary to start this bundle. This method can be used to register services or to allocate any resources that this bundle needs.

This method must complete and return to its caller in a timely manner.

Parameters: context - The execution context of the bundle being started.

Throws: java.lang.Exception - If this method throws an exception, this bundle is marked as stopped and the Framework will remove this bundle's listeners, unregister all services registered by this bundle, and release all services used by this bundle.

See Also: start()

public void **stop**(BundleContext context)
    throws java.lang.Exception

Called when this bundle is stopped so the Framework can perform the bundle-specific activities necessary to stop the bundle. In general, this method should undo the work that the BundleActivator.start method started. There should be no active threads when this bundle returns. A stopped bundle should be stopped and should not call any Framework objects.

This method must complete and return to its caller in a timely manner.

Parameters: context - The execution context of the bundle being stopped.

Throws: java.lang.Exception - If this method throws an exception, the bundle is still marked as stopped, and the Framework will remove the bundle's listeners, unregister all services registered by the bundle, and release all services used by the bundle.

See Also: stop()

2.21.4 BundleContext

public interface BundleContext

A bundle's execution context within the Framework. The context is used to grant access to other methods so that this bundle can interact with the Framework.

BundleContext methods allow a bundle to:

- Subscribe to events published by the Framework.
- Register services in the Framework service registry.
- Retrieve ServiceReferences from the Framework service registry.
- Get and release service objects for a referenced service.
- Install new bundles in the Framework.
- Get the list of bundles installed in the Framework.
- Get the Bundle object for a bundle.
- Create File objects for files in a persistent storage area provided for the bundle by the Framework.

A BundleContext object will be created and provided to this bundle when it is started using the start(BundleContext) method. The same BundleContext object will be passed to this bundle when it is stopped using the stop(BundleContext) method. BundleContext is generally for the private use of this bundle and is not meant to be shared with other bundles in the OSGi environment. BundleContext is used when resolving ServiceListeners and EventListeners.

The BundleContext object is only valid during an execution instance of this bundle; that is, during the period from when this bundle is called by BundleActivator.start until after this bundle is called and returns from BundleActivator.stop (or if BundleActivator.start terminates with an exception). If the BundleContext object is used subsequently, an IllegalStateException may be thrown. When this bundle is restarted, a new BundleContext object will be created.

The Framework is the only entity that can create BundleContext objects and they are only valid within the Framework that created them.

Note: A single virtual machine may host multiple Framework instances at any given time, but objects created by one Framework instance cannot be used by bundles running in the execution context of another Framework instance.

2.21.4.1   **Methods**

public void **addBundleListener**(BundleListener listener)
   throws java.lang.IllegalStateException

Adds the specified BundleListener object to this context bundle's list of listeners if not already present. See getBundle() for a definition of context bundle. BundleListener objects are notified when a bundle has a lifecycle state change.

If this context bundle's list of listeners already contains a listener l such that (l==listener), this method does nothing.

Parameters: listener - The BundleListener to be added.

Throws: java.lang.IllegalStateException - If this context bundle has stopped.

See Also: BundleEvent, BundleListener

public void **addFrameworkListener**(FrameworkListener listener)
   throws java.lang.IllegalStateException

Adds the specified FrameworkListener object to this context bundle's list of listeners if not already present. See getBundle() for a definition of context bundle. FrameworkListeners are notified of general Framework events.

If this context bundle's list of listeners already contains a listener l such that (l==listener), this method does nothing.

Parameters: listener - The FrameworkListener object to be added.

Throws: java.lang.IllegalStateException - If this context bundle has stopped.

See Also: FrameworkEvent, FrameworkListener

public void **addServiceListener**(ServiceListener listener)
   throws java.lang.IllegalStateException

Adds the specified ServiceListener object to this context bundle's list of listeners.

This method is the same as calling BundleContext.addService-Listener(ServiceListener listener, String filter) with filter set to null.

Parameters: listener - The ServiceListener object to be added.

Throws: java.lang.IllegalStateException - If this context bundle has stopped.

See Also: addServiceListener(ServiceListener, String)

public void **addServiceListener**(ServiceListener listener,
    java.lang.String filter)
    throws InvalidSyntaxException, java.lang.IllegalStateException

Adds the specified ServiceListener object with the specified filter to this context bundle's list of listeners.

See getBundle() for a definition of context bundle, and Filter for a description of the filter syntax. ServiceListener objects are notified when a service has a lifecycle state change.

If this context bundle's list of listeners already contains a listener l such that (l==listener), this method replaces that listener's filter (which may be null) with the specified one (which may be null).

The listener is called if the filter criteria is met. To filter based upon the class of the service, the filter should reference the OBJECTCLASS property. If filter is null, all services are considered to match the filter.

If the Java Runtime Environment supports permissions, the ServiceListener object will be notified of a service event only if the bundle that is registering it has the ServicePermission to get the service using at least one of the named classes the service was registered under.

Parameters: listener - The ServiceListener object to be added.

filter - The filter criteria.

Throws: InvalidSyntaxException - If filter contains an invalid filter string which cannot be parsed.

java.lang.IllegalStateException - If this context bundle has stopped.

See Also: ServiceEvent, ServiceListener, ServicePermission

public Filter **createFilter**(java.lang.String filter)
    throws InvalidSyntaxException

Creates a Filter object. This Filter object may be used to match a ServiceReference object or a Dictionary object. See Filter for a description of the filter string syntax.

If the filter cannot be parsed, an InvalidSyntaxException will be thrown with a human readable message where the filter became unparsable.

Parameters: filter - The filter string.

Returns: A Filter object encapsulating the filter string.

Throws:	InvalidSyntaxException - If filter contains an invalid filter string that cannot be parsed.
Since:	1.1

public Bundle **getBundle**()
   throws java.lang.IllegalStateException

Returns the Bundle object for this context bundle.

The context bundle is defined as the bundle that was assigned this BundleContext in its BundleActivator.

Returns:	The context bundle's Bundle object.
Throws:	java.lang.IllegalStateException - If this context bundle has stopped.

public Bundle **getBundle**(long id)

Returns the bundle with the specified identifier.

Parameters:	id - The identifier of the bundle to retrieve.
Returns:	A Bundle object, or null if the identifier does not match any installed bundle.

public Bundle[] **getBundles**()

Returns a list of all installed bundles.

This method returns a list of all bundles installed in the OSGi environment at the time of the call to this method. However, as the Framework is a very dynamic environment, bundles can be installed or uninstalled at anytime.

Returns:	An array of Bundle objects; one object per installed bundle.

public java.io.File **getDataFile**(java.lang.String filename)
   throws java.lang.IllegalStateException

Creates a File object for a file in the persistent storage area provided for the bundle by the Framework. This method will return null if the platform does not have file system support.

A File object for the base directory of the persistent storage area provided for the context bundle by the Framework can be obtained by calling this method with an empty string (" ") as filename. See get-Bundle() for a definition of context bundle.

If the Java Runtime Environment supports permissions, the Framework will ensure that the bundle has the java.io.FilePermission with actions read, write, execute, delete for all files (recursively) in the persistent storage area provided for the context bundle.

Parameters: filename - A relative name to the file to be accessed.

Returns: A File object that represents the requested file or null if the platform does not have file system support.

Throws: java.lang.IllegalStateException - If the context bundle has stopped.

public java.lang.String **getProperty**(java.lang.String key)

Returns the value of the specified environment property.

The following standard property keys are valid:

FRAMEWORK_VERSION

The OSGi Framework version.

FRAMEWORK_VENDOR

The Framework implementation vendor.

FRAMEWORK_LANGUAGE

The language being used. See ISO 639 for possible values.

FRAMEWORK_OS_NAME

The host computer operating system.

FRAMEWORK_OS_VERSION

The host computer operating system version number.

FRAMEWORK_PROCESSOR

The host computer processor name.

Note: The last four standard properties are used by the BUNDLE_NATIVECODE Manifest header's matching algorithm for selecting native language code.

Parameters: key - The name of the requested property.

Returns: The value of the requested property, or null if the property is undefined.

public java.lang.Object **getService**(ServiceReference reference)
    throws java.lang.SecurityException,
    java.lang.IllegalStateException

Returns the specified service object for a service.

A bundle's use of a service is tracked by the bundle's use count of that service. Each time a service's service object is returned by get-Service(ServiceReference) the context bundle's use count for that service is incremented by one. Each time the service is released by ungetService(ServiceReference) the context bundle's use count for that service is decremented by one.

When a bundle's use count for a service drops to zero, the bundle should no longer use that service. See getBundle() for a definition of context bundle.

This method will always return null when the service associated with this reference has been unregistered.

The following steps are required to get the service object:

1. If the service has been unregistered, null is returned.

2. The context bundle's use count for this service is incremented by one.

3. If the context bundle's use count for the service is currently one and the service was registered with an object implementing the ServiceFactory interface, the getService(Bundle, ServiceRegistration) method is called to create a service object for the context bundle. This service object is cached by the Framework. While the context bundle's use count for the service is greater than zero, subsequent calls to get the services's service object for the context bundle will return the cached service object.
If the service object returned by the ServiceFactory object is not an instanceof all the classes named when the service was registered or the ServiceFactory object throws an exception, null is returned and a Framework event of type ERROR is broadcast.

4. The service object for the service is returned.

Parameters: reference - A reference to target service object's service.

Returns: A service object for the service associated with reference, or null if the service is not registered.

Throws: java.lang.SecurityException - If the caller does not have the ServicePermission to get the service using at least one of the named classes the service was registered under, and the Java Runtime Environment supports permissions.

                     java.lang.IllegalStateException - If the context bundle has stopped.

See Also: ungetService(ServiceReference), ServiceFactory

public ServiceReference **getServiceReference**(java.lang.String
   clazz)

Returns a ServiceReference object for a service that implements,
and was registered under, the specified class.

This ServiceReference object is valid at the time of the call to this
method, however as the Framework is a very dynamic environment,
services can be modified or unregistered at anytime.

This method is the same as calling getServiceReferences(String,
String) with a null filter string. It is provided as a convenience for
when the caller is interested in any service that implements the
specified class.

If multiple such services exist, the service with the highest ranking (
as specified in its SERVICE_RANKING property) is returned.

If there is a tie in ranking, the service with the lowest service ID (as
specified in its SERVICE_ID property); that is, the service that was
registered first is returned.

Parameters: clazz - The class name with which the service was registered.

Returns: A ServiceReference object, or null if no services are registered which
                     implement the named class.

See Also: getServiceReferences(String, String)

public ServiceReference[] **getServiceReferences**(java.lang.String
   clazz, java.lang.String filter)
   throws InvalidSyntaxException

Returns a list of ServiceReference objects. This method returns a list
of ServiceReference objects for services which implement and were
registered under the specified class and match the specified filter criteria.

The list is valid at the time of the call to this method, however as the
Framework is a very dynamic environment, services can be modified or unregistered at anytime.

filter is used to select the registered service whose properties objects
contain keys and values which satisfy the filter. See Filter for a
description of the filter string syntax.

If filter is null, all registered services are considered to match the filter.

If filter cannot be parsed, an InvalidSyntaxException will be thrown
with a human readable message where the filter became unparsable.

The following steps are required to select a service:

1. If the Java Runtime Environment supports permissions, the caller is checked for the ServicePermission to get the service with the specified class. If the caller does not have the correct permission, null is returned.

2. If the filter string is not null, the filter string is parsed and the set of registered services which satisfy the filter is produced. If the filter string is null, then all registered services are considered to satisfy the filter.

3. If clazz is not null, the set is further reduced to those services which are an instanceof and were registered under the specified class. The complete list of classes of which a service is an instance and which were specified when the service was registered is available from the service's OBJECTCLASS property.

4. An array of ServiceReference to the selected services is returned.

Parameters: clazz - The class name with which the service was registered, or null for all services.

filter - The filter criteria.

Returns: An array of ServiceReference objects, or null if no services are registered which satisfy the search.

Throws: InvalidSyntaxException - If filter contains an invalid filter string which cannot be parsed.

public Bundle **installBundle**(java.lang.String location)
    throws BundleException, java.lang.SecurityException

Installs the bundle from the specified location string. A bundle is obtained from location as interpreted by the Framework in an implementation dependent manner.

Every installed bundle is uniquely identified by its location string, typically in the form of a URL.

The following steps are required to install a bundle:

1. If a bundle containing the same location string is already installed, the Bundle object for that bundle is returned.

2. The bundle's content is read from the location string. If this fails, a BundleException is thrown.

3. The bundle's Bundle-ClassPath and Bundle-NativeCode dependencies are resolved. If this fails, a BundleException is thrown.

4. The bundle's associated resources are allocated. The associated resources minimally consist of a unique identifier, and a persistent storage area if the platform has file system support. If this step fails, a BundleException is thrown.

5. The bundle's state is set to INSTALLED.

6. If the bundle has not declared an Import-Package Manifest header (that is, the bundle does not depend on any packages from other OSGi bundles), the bundle's state may be set to RESOLVED.

7. A bundle event of type INSTALLED is broadcast.

8. The Bundle object for the newly installed bundle is returned.

Postconditions, no exceptions thrown

- getState() in {INSTALLED}, RESOLVED}.
- Bundle has a unique ID.

Postconditions, when an exception is thrown

- Bundle is not installed and no trace of the bundle exists.

Parameters: location - The location identifier of the bundle to install.

Returns: The Bundle object of the installed bundle.

Throws: BundleException - If the installation failed.

java.lang.SecurityException - If the caller does not have the appropriate AdminPermission, and the Java Runtime Environment supports permissions.

public Bundle **installBundle**(java.lang.String location,
    java.io.InputStream in)
    throws BundleException

Installs the bundle from the specified InputStream object.

This method performs all of the steps listed in Bundle-Context.installBundle(String location), except that the bundle's content will be read from the InputStream object. The location identifier string specified will be used as the identity of the bundle.

This method will always close the InputStream object, even if an exception is thrown.

Parameters: location - The location identifier of the bundle to install.

in - The InputStream object from which this bundle will be read.

Returns: The Bundle object of the installed bundle.

Throws: BundleException - If the provided stream cannot be read.

See Also: installBundle(String)

public ServiceRegistration **registerService**(java.lang.String[]
    clazzes, java.lang.Object service, java.util.Dictionary properties)
    throws java.lang.IllegalArgumentException,
    java.lang.SecurityException, java.lang.IllegalStateException

Registers the specified service object with the specified properties under the specified class names into the Framework. A Service-Registration object is returned. The ServiceRegistration object is for the private use of the bundle registering the service and should not be shared with other bundles. The registering bundle is defined to be the context bundle. See getBundle() for a definition of context bundle. Other bundles can locate the service by using either the getServiceReferences(String, String) or getServiceReference(String) method.

A bundle can register a service object that implements the ServiceFactory interface to have more flexibility in providing service objects to other bundles.

The following steps are required to register a service:

1. If service is not a ServiceFactory, an IllegalArgumentException is thrown if service is not an instanceof all the classes named.

2. The Framework adds these service properties to the specified Dictionary (which may be null): a property named SERVICE_ID identifying the registration number of the service, and a property named OBJECTCLASS containing all the specified classes. If any of these properties have already been specified by the registering bundle, their values will be overwritten by the Framework.

3. The service is added to the Framework service registry and may now be used by other bundles.

4. A service event of type REGISTERED is synchronously sent.

5. A ServiceRegistration object for this registration is returned.

Parameters: clazzes - The class names under which the service can be located. The class names in this array will be stored in the service's properties under the key OBJECTCLASS .

service - The service object or a ServiceFactory object.

Returns:

properties - The properties for this service. The keys in the properties object must all be Strings. See Constants for a list of standard service property keys. Changes should not be made to this object after calling this method. To update the service's properties the setProperties(Dictionary) method must be called. properties may be null if the service has no properties.

Returns: A ServiceRegistration object for use by the bundle registering the service to update the service's properties or to unregister the service.

Throws: java.lang.IllegalArgumentException - If one of the following is true:

- service is null.
- service is not a ServiceFactory object and is not an instanceof all the named classes in clazzes.
- properties contains case variants of the same key name.

java.lang.SecurityException - If the caller does not have the ServicePermission to register the service for all the named classes and the Java Runtime Environment supports permissions.

java.lang.IllegalStateException - If this context bundle has stopped.

See Also: ServiceRegistration, ServiceFactory

public ServiceRegistration **registerService**(java.lang.String clazz, java.lang.Object service, java.util.Dictionary properties)

Registers the specified service object with the specified properties under the specified class name into the Framework.

This method is otherwise identical to registerService(String[], Object, Dictionary) and is provided as a convenience when service will only be registered under a single class name. Note that even in this case the value of the service's OBJECTCLASS property will be an array of strings, rather than just a single string.

See Also: registerService(String[], Object, Dictionary)

public void **removeBundleListener**(BundleListener listener)
    throws java.lang.IllegalStateException

Removes the specified BundleListener object from this context bundle's list of listeners. See getBundle() for a definition of context bundle.

If listener is not contained in this context bundle's list of listeners, this method does nothing.

Parameters: listener - The BundleListener object to be removed.

Throws: java.lang.IllegalStateException - If this context bundle has stopped.

public void **removeFrameworkListener**(FrameworkListener
    listener)
    throws java.lang.IllegalStateException

Removes the specified FrameworkListener object from this context bundle's list of listeners. See getBundle() for a definition of context bundle.

If listener is not contained in this context bundle's list of listeners, this method does nothing.

Parameters: listener - The FrameworkListener object to be removed.

Throws: java.lang.IllegalStateException - If this context bundle has stopped.

public void **removeServiceListener**(ServiceListener listener)
    throws java.lang.IllegalStateException

Removes the specified ServiceListener object from this context bundle's list of listeners. See getBundle() for a definition of context bundle.

If listener is not contained in this context bundle's list of listeners, this method does nothing.

Parameters: listener - The ServiceListener to be removed.

Throws: java.lang.IllegalStateException - If this context bundle has stopped.

public boolean **ungetService**(ServiceReference reference)
    throws java.lang.IllegalStateException

Releases the service object referenced by the specified ServiceReference object. If the context bundle's use count for the service is zero, this method returns false. Otherwise, the context bundle's use count for the service is decremented by one. See getBundle() for a definition of context bundle.

The service's service object should no longer be used and all references to it should be destroyed when a bundle's use count for the service drops to zero.

The following steps are required to unget the service object:

1. If the context bundle's use count for the service is zero or the service has been unregistered, false is returned.

2. The context bundle's use count for this service is decremented by one.

3. If the context bundle's use count for the service is currently zero and the service was registered with a ServiceFactory object, the

ungetService(Bundle, ServiceRegistration, Object) method is called to release the service object for the context bundle.

4. true is returned.

Parameters: reference - A reference to the service to be released.

Returns: false if the context bundle's use count for the service is zero or if the service has been unregistered; true otherwise.

Throws: java.lang.IllegalStateException - If the context bundle has stopped.

See Also: getService(ServiceReference), ServiceFactory

2.21.5 BundleEvent

public class BundleEvent extends java.util.EventObject

All Implement-ed Interfaces: java.io.Serializable

A Framework event describing a bundle lifecycle change.

BundleEvent objects are delivered to BundleListener objects when a change occurs in a bundle's lifecycle. A type code is used to identify the event type for future extendability.

OSGi reserves the right to extend the set of types.

2.21.5.1 Fields

public static final int **INSTALLED**

This bundle has been installed.

The value of INSTALLED is 0x00000001.

See Also: installBundle(String)

public static final int **STARTED**

This bundle has been started.

The value of STARTED is 0x00000002.

See Also: start()

public static final int **STOPPED**

This bundle has been stopped.

The value of STOPPED is 0x00000004.

See Also: stop()

public static final int **UNINSTALLED**

This bundle has been uninstalled.

The value of UNINSTALLED is 0x00000010.

See Also: uninstall()

public static final int **UPDATED**

This bundle has been updated.

The value of UPDATED is 0x00000008.

See Also: update()

2.21.5.2   **Constructors**
public **BundleEvent**(int type, Bundle bundle)

Creates a bundle event of the specified type.

Parameters: type - The event type.

bundle - The bundle which had a lifecycle change.

2.21.5.3   **Methods**
public Bundle **getBundle**()

Returns the bundle which had a lifecycle change. This bundle is the source of the event.

Returns: A bundle that had a change occur in its lifecycle.

public int **getType**()

Returns the type of lifecyle event. The type values are:

- INSTALLED
- STARTED
- STOPPED
- UPDATED
- UNINSTALLED

Returns: The type of lifecycle event.

2.21.6   **BundleException**

public class BundleException extends java.lang.Exception

All Implement-   java.io.Serializable
ed Interfaces:
                   A Framework exception used to indicate that a bundle lifecycle problem occurred.

Bundle Exception object is created by the Framework to denote an exception condition in the lifecycle of a bundle. Bundle Exceptions should not be created by bundle developers.

2.21.6.1	**Constructors**

public **BundleException**(java.lang.String msg)

Creates a Bundle Exception object with the specified message.

Parameters: msg - The message.

public **BundleException**(java.lang.String msg, java.lang.Throwable
throwable)

Creates a Bundle Exception that wraps another exception.

Parameters: msg - The associated message.

throwable - The nested exception.

2.21.6.2   **Methods**

public java.lang.Throwable **getNestedException**()

Returns any nested exceptions included in this exception.

Returns: The nested exception; null if there is no nested exception.

2.21.7   **BundleListener**

public interface BundleListener extends java.util.EventListener

All Known Sub-interfaces: SynchronousBundleListener

All Superinter-faces: java.util.EventListener

A Bundle Event listener.

Bundle Listener is a listener interface that may be implemented by a bundle developer.

A Bundle Listener object is registered with the Framework using the addBundleListener(BundleListener) method. Bundle Listeners are called with a Bundle Event object when a bundle has been installed, started, stopped, updated, or uninstalled.

See Also: BundleEvent

2.21.7.1   **Methods**

public void **bundleChanged**(BundleEvent event)

Receives notification that a bundle has had a lifecycle change.

Parameters: event - The Bundle Event.

2.21.8   **Configurable**

public interface Configurable

Supports a configuration object.

Configurable is an interface that should be used by a bundle developer in support of a configurable service. Bundles that need to configure a service may test to determine if the service object is an instanceof Configurable.

2.21.8.1   **Methods**

public java.lang.Object **getConfigurationObject**()
    throws java.lang.SecurityException

Returns this service's configuration object.

Services implementing Configurable should take care when returning a service configuration object since this object is probably sensitive.

If the Java Runtime Environment supports permissions, it is recommended that the caller is checked for the appropriate permission before returning the configuration object. It is recommended that callers possessing the appropriate AdminPermission always be allowed to get the configuration object.

Returns: The configuration object for this service.

Throws: java.lang.SecurityException - If the caller does not have an appropriate permission and the Java Runtime Environment supports permissions.

2.21.9   # Constants

public interface Constants

Defines standard names for the OSGi environment property, service property, and Manifest header attribute keys.

The values associated with these keys are of type java.lang.String, unless otherwise indicated.

Since: 1.1

See Also: getHeaders(), getProperty(String), registerService(String[], Object, Dictionary)

2.21.9.1   **Fields**

public static final java.lang.String **BUNDLE_ACTIVATOR**

Manifest header attribute (named "Bundle-Activator") identifying the bundle's activator class.

If present, this header specifies the name of the bundle resource class that implements the BundleActivator interface and whose start and stop methods are called by the Framework when the bundle is started and stopped, respectively.

The attribute value may be retrieved from the Dictionary object returned by the Bundle.getHeaders method.

public static final java.lang.String **BUNDLE_CATEGORY**

Manifest header (named "Bundle-Category") identifying the bundle's category.

The attribute value may be retrieved from the Dictionary object returned by the Bundle.getHeaders method.

public static final java.lang.String **BUNDLE_CLASSPATH**

Manifest header (named "Bundle-ClassPath") identifying a list of nested JAR files, which are bundle resources used to extend the bundle's classpath.

The attribute value may be retrieved from the Dictionary object returned by the Bundle.getHeaders method.

public static final java.lang.String **BUNDLE_CONTACTADDRESS**

Manifest header (named "Bundle-ContactAddress") identifying the contact address where problems with the bundle may be reported; for example, an email address.

The attribute value may be retrieved from the Dictionary object returned by the Bundle.getHeaders method.

public static final java.lang.String **BUNDLE_COPYRIGHT**

Manifest header (named "Bundle-Copyright") identifying the bundle's copyright information, which may be retrieved from the Dictionary object returned by the Bundle.getHeaders method.

public static final java.lang.String **BUNDLE_DESCRIPTION**

Manifest header (named "Bundle-Description") containing a brief description of the bundle's functionality.

The attribute value may be retrieved from the Dictionary object returned by the Bundle.getHeaders method.

public static final java.lang.String **BUNDLE_DOCURL**

Manifest header (named "Bundle-DocURL") identifying the bundle's documentation URL, from which further information about the bundle may be obtained.

The attribute value may be retrieved from the Dictionary object returned by the Bundle.getHeaders method.

public static final java.lang.String **BUNDLE_NAME**

Manifest header (named "Bundle-Name") identifying the bundle's name.

The attribute value may be retrieved from the Dictionary object returned by the Bundle.getHeaders method.

public static final java.lang.String **BUNDLE_NATIVECODE**

Manifest header (named "Bundle-NativeCode") identifying a number of hardware environments and the native language code libraries that the bundle is carrying for each of these environments.

The attribute value may be retrieved from the Dictionary object returned by the Bundle.getHeaders method.

public static final java.lang.String
   BUNDLE_NATIVECODE_LANGUAGE

Manifest header attribute (named "language") identifying the language in which the native bundle code is written specified in the Bundle-NativeCode Manifest header). see ISO 639 for possible values).

The attribute value is encoded in the Bundle-NativeCode Manifest header like:

```
Bundle-NativeCode: http.so ; language=nl_be ...
```

public static final java.lang.String **BUNDLE_NATIVECODE_OSNAME**

Manifest header attribute (named "osname") identifying the operating system required to run native bundle code specified in the Bundle-NativeCode Manifest header).

The attribute value is encoded in the Bundle-NativeCode Manifest header like:

```
Bundle-NativeCode: http.so ; osname=Linux ...
```

public static final java.lang.String
  BUNDLE_NATIVECODE_OSVERSION

Manifest header attribute (named "osversion") identifying the operating system version required to run native bundle code specified in the Bundle-NativeCode Manifest header).

The attribute value is encoded in the Bundle-NativeCode Manifest header like:

```
Bundle-NativeCode: http.so ; osversion="2.34" ...
```

public static final java.lang.String
  BUNDLE_NATIVECODE_PROCESSOR

Manifest header attribute (named "processor") identifying the processor required to run native bundle code specified in the Bundle-NativeCode Manifest header).

The attribute value is encoded in the Bundle-NativeCode Manifest header like:

```
Bundle-NativeCode: http.so ; processor=x86 ...
```

public static final java.lang.String **BUNDLE_UPDATELOCATION**

Manifest header (named "Bundle-UpdateLocation") identifying the location from which a new bundle version is obtained during a bundle update operation.

The attribute value may be retrieved from the Dictionary object returned by the Bundle.getHeaders method.

public static final java.lang.String **BUNDLE_VENDOR**

Manifest header (named "Bundle-Vendor") identifying the bundle's vendor.

The attribute value may be retrieved from the Dictionary object returned by the Bundle.getHeaders method.

public static final java.lang.String **BUNDLE_VERSION**

Manifest header (named "Bundle-Version") identifying the bundle's version.

The attribute value may be retrieved from the Dictionary object returned by the Bundle.getHeaders method.

public static final java.lang.String **EXPORT_PACKAGE**

Manifest header (named "Export-Package") identifying the names (and optionally version numbers) of the packages that the bundle offers to the Framework for export.

The attribute value may be retrieved from the Dictionary object returned by the Bundle.getHeaders method.

public static final java.lang.String **EXPORT_SERVICE**

Manifest header (named "Export-Service") identifying the fully qualified class names of the services that the bundle may register (used for informational purposes only).

The attribute value may be retrieved from the Dictionary object returned by the Bundle.getHeaders method.

public static final java.lang.String **FRAMEWORK_LANGUAGE**

Framework environment property (named "org.osgi.framework.language") identifying the Framework implementation language (see ISO 639 for possible values).

The value of this property may be retrieved by calling the Bundle-Context.getProperty method.

public static final java.lang.String **FRAMEWORK_OS_NAME**

Framework environment property (named "org.osgi.framework.os.name") identifying the Framework host-computer's operating system.

The value of this property may be retrieved by calling the Bundle-Context.getProperty method.

public static final java.lang.String **FRAMEWORK_OS_VERSION**

Framework environment property (named "org.osgi.framework.os.version") identifying the Framework host-computer's operating system version number.

The value of this property may be retrieved by calling the Bundle-Context.getProperty method.

public static final java.lang.String **FRAMEWORK_PROCESSOR**

Framework environment property (named "org.osgi.framework.processor") identifying the Framework host-computer's processor name.

The value of this property may be retrieved by calling the Bundle-Context.getProperty method.

public static final java.lang.String **FRAMEWORK_VENDOR**

Framework environment property (named "org.osgi.framework.vendor") identifying the Framework implementation vendor.

The value of this property may be retrieved by calling the Bundle-Context.getProperty method.

public static final java.lang.String **FRAMEWORK_VERSION**

Framework environment property (named "org.osgi.framework.version") identifying the Framework version.

The value of this property may be retrieved by calling the Bundle-Context.getProperty method.

public static final java.lang.String **IMPORT_PACKAGE**

Manifest header (named "Import-Package") identifying the names (and optionally, version numbers) of the packages that the bundle is dependent on.

The attribute value may be retrieved from the Dictionary object returned by the Bundle.getHeaders method.

public static final java.lang.String **IMPORT_SERVICE**

Manifest header (named "Import-Service") identifying the fully qualified class names of the services that the bundle requires (used for informational purposes only).

The attribute value may be retrieved from the Dictionary object returned by the Bundle.getHeaders method.

public static final java.lang.String **OBJECTCLASS**

Service property (named "objectClass") identifying all of the class names under which a service was registered in the Framework.

This property is set by the Framework when a service is registered.

public static final java.lang.String **PACKAGE_SPECIFICATION_VERSION**

Manifest header attribute (named "specification-version") identifying the version of a package specified in the Export-Package or Import-Package Manifest header.

The attribute value is encoded in the Export-Package or Import-Package Manifest header like:

```
Import-Package: org.osgi.framework ; specification-
version="1.1"
```

public static final java.lang.String **SERVICE_DESCRIPTION**

Service property (named "service.description") identifying a service's description.

This property may be supplied in the properties Dictionary object passed to the BundleContext.registerService method.

public static final java.lang.String **SERVICE_ID**

Service property (named "service.id") identifying a service's registration number (of type java.lang.Long).

The value of this property is assigned by the Framework when a service is registered. The Framework assigns a unique value that is larger than all previously assigned values since the Framework was started. These values are NOT persistent across restarts of the Framework.

public static final java.lang.String **SERVICE_PID**

Service property (named "service.pid") identifying a service's persistent identifier.

This property may be supplied in the properties Dictionary object passed to the BundleContext.registerService method.

A service's persistent identifier uniquely identifies the service and persists across multiple Framework invocations.

By convention, every bundle has its own unique namespace, starting with the bundle's identifier (see getBundleId()) and followed by a dot (.). A bundle may use this as the prefix of the persistent identifiers for the services it registers.

public static final java.lang.String **SERVICE_RANKING**

Service property (named "service.ranking") identifying a service's ranking number (of type java.lang.Integer).

This property may be supplied in the properties Dictionary object passed to the BundleContext.registerService method.

The service ranking is used by the Framework to determine the *default* service to be returned from a call to the getServiceReference(String) method: If more than one service implements the specified class, the ServiceReference object with the highest ranking is returned.

The default ranking is 0. A service with a ranking of Integer.MAX_VALUE is very likely to be returned as the default service, whereas a service with a ranking of Integer.MIN_VALUE is very unlikely to be returned.

If the supplied property value is not of type java.lang.Integer, it is deemed to have a ranking value of 0.

public static final java.lang.String **SERVICE_VENDOR**

Service property (named "service.vendor") identifying a service's vendor.

This property may be supplied in the properties Dictionary object passed to the BundleContext.registerService method.

public static final java.lang.String **SYSTEM_BUNDLE_LOCATION**

Location identifier of the OSGi *system bundle*, which is defined to be "System Bundle".

2.21.10   Filter

public interface Filter

An RFC 1960-based Filter.

Filter objects can be created by calling createFilter(String) with the chosen filter string.

A Filter object can be used numerous times to determine if the match argument matches the filter string that was used to create the Filter object.

The syntax of a filter string is the string representation of LDAP search filters as defined in RFC 1960: *A String Representation of LDAP Search Filters* (available at http://www.ietf.org/rfc/rfc1960.txt). It should be noted that RFC 2254: *A String Representation of LDAP Search Filters* (available at http://www.ietf.org/rfc/rfc2254.txt) supercedes RFC 1960 but only adds extensible matching and is not applicable for this OSGi Framework API.

The string representation of an LDAP search filter uses a prefix format, and is defined with the following grammar.

```
<filter> ::= '(' <filtercomp> ')'
```

```
<filtercomp> ::= <and> | <or> | <not> | <item>
<and> ::= '&' <filterlist>
<or> ::= '|' <filterlist>
<not> ::= '!' <filter>
<filterlist> ::= <filter> | <filter> <filterlist>
<item> ::= <simple> | <present> | <substring>
<simple> ::= <attr> <filtertype> <value>
<filtertype> ::= <equal> | <approx> | <greater> |
<less>
<equal> ::= '='
<approx> ::= '~='
<greater> ::= '>='
<less> ::= '<='
<present> ::= <attr> '=*'
<substring> ::= <attr> '=' <initial> <any> <final>
<initial> ::= NULL | <value>
<any> ::= '*' <starval>
<starval> ::= NULL | <value> '*' <starval>
<final> ::= NULL | <value>
```

<attr> is a string representing an attribute, or key, in the properties
objects of the services registered in the Framework. Attribute names
are not case sensitive; that is, cn and CN both refer to the same
attribute. <attr> should contain no spaces though white space is
allowed between the initial parenthesis "(" and the start of the key,
and between the end of the key and the equal sign "=". <value> is a
string representing the value, or part of one, of a key in the properties objects of the registered services. If a <value> must contain one
of the characters '*' or '(' or ')', these characters should be escaped by
preceding them with the backslash '\' character. Spaces are significant in <value>. Space charactes are defined by
java.lang.Character.isWhiteSpace(). Note that although both the
<substring> and <present> productions can produce the 'attr=*'
construct; this construct is used only to denote a presence filter.

Examples of LDAP filters are:

```
"(cn=Babs Jensen)"
"(!(cn=Tim Howes))"
"(&(" + Constants.OBJECTCLASS + "=Person)(|(
sn=Jensen)(cn=Babs J*)))"
"(o=univ*of*mich*)"
```

The approximate match (~=) is implementation specific but should
at least ignore case and white space differences. Optional are codes
like soundex or other smart "closeness" comparisons.

Comparison of values is not straightforward. Strings are compared differently than numbers and it is possible for a key to have multiple values. Note that keys in the match argument must always be strings. The comparison is defined by the object type of the key's value. The following rules apply for comparison:

```
Property Value Type     Comparison Type
String        String comparison
Integer, Long, Float,
Double, Byte, Short,
BigInteger,BigDecimal  Numerical comparison
Character   Character comparison
Boolean     Equality comparisons only
[ ] (array)    Recursively applied to values
Vector      Recursively applied to elements
```

Arrays of primitives are also supported. A filter matches a key that has multiple values if it matches at least one of those values. For example,

```
Dictionary d = new Hashtable();
d.put( "cn", new String[] { "a", "b", "c" } );
```

d will match (cn=a) and also (cn=b)

A filter component that references a key having an unrecognizable data type will evaluate to false.

Since: 1.1

2.21.10.1   **Methods**

public boolean **equals**(java.lang.Object obj)

Compares this Filter object to another object.

Overrides: java.lang.Object.equals(java.lang.Object) in class java.lang.Object

Parameters: obj - The object to compare against this Filter object.

Returns: If the other object is a Filter object, then returns this.to String( ).equals(obj.to String(); false otherwise.

public int **hashCode**()

Returns the hashCode for this Filter object.

Overrides: java.lang.Object.hashCode() in class java.lang.Object

Returns: The hashCode of the filter string; that is, this.to String().hash Code( ).

```
public boolean match(java.util.Dictionary dictionary)
    throws IllegalArgumentException
```

Filter using a Dictionary object. The Filter is executed using the Dictionary object's keys and values.

Parameters: dictionary - The Dictionary object whose keys are used in the match.

Returns: true if the Dictionary object's keys and values match this filter; false otherwise.

Throws: IllegalArgumentException - If dictionary contains case variants of the same key name.

```
public boolean match(ServiceReference reference)
```

Filter using a service's properties.

The filter is executed using properties of the referenced service.

Parameters: reference - The reference to the service whose properties are used in the match.

Returns: true if the service's properties match this filter; false otherwise.

```
public java.lang.String toString()
```

Returns this Filter object's filter string.

The filter string is normalized by removing whitespace which does not affect the meaning of the filter.

Overrides: java.lang.Object.toString() in class java.lang.Object

Returns: Filter string.

2.21.11   FrameworkEvent

public class FrameworkEvent extends java.util.EventObject

All Implement-   java.io.Serializable
ed Interfaces:
A general Framework event.

FrameworkEvent is the event class used when notifying listeners of general events occuring within the OSGI environment. A type code is used to identify the event type for future extendability.

OSGi reserves the right to extend the set of event types.

2.21.11.1   Fields
public static final int **ERROR**

An error has occurred.

There was an error associated with a bundle.

The value of ERROR is 0x00000002.

public static final int **STARTED**

The Framework has started.

This event is broadcast when the Framework has started after all installed bundles that are marked to be started have been started.

The value of STARTED is 0x00000001.

2.21.11.2   **Constructors**

public **FrameworkEvent**(int type, Bundle bundle,
    java.lang.Throwable throwable)

Creates a Framework event with a related bundle and exception.

This constructor is used for Framework events of type ERROR.

Parameters: type - The event type.

bundle - The related bundle.

throwable - The related exception.

public **FrameworkEvent**(int type, java.lang.Object source)

Creates a Framework event.

This constructor is used for Framework events of type STARTED.

Parameters: type - The event type.

source - The event source object. This may not be null.

2.21.11.3   **Methods**

public Bundle **getBundle**()

Returns the bundle associated with the event.

If the event type is ERROR, this metod returns the bundle related to the error. This bundle is also the source of the event.

Returns: A bundle if an event of type ERROR or null.

public java.lang.Throwable **getThrowable**()

Returns the exception associated with the event.

If the event type is ERROR, this method returns the exception related to the error.

Returns: An exception if an event of type ERROR or null.

public int **getType**()

Returns the type of bundle state change.

The type values are:

- STARTED
- ERROR

Returns: The type of state change.

2.21.12   FrameworkListener

public interface FrameworkListener extends
**   java.util.EventListener**

All Superinter-   java.util.EventListener
faces:
                    A FrameworkEvent listener.

FrameworkListener is a listener interface that may be implemented
by a bundle developer. A FrameworkListener object is registered
with the Framework using the addFrameworkListener(
FrameworkListener) method. FrameworkListener objects are called
with a FrameworkEvent objects when the Framework starts and
when asynchronous errors occur.

See Also: FrameworkEvent

2.21.12.1   Methods
public void **frameworkEvent**(FrameworkEvent event)

Receives notification of a general FrameworkEvent object.

Parameters: event - The FrameworkEvent object.

2.21.13   InvalidSyntaxException

public class InvalidSyntaxException extends java.lang.Exception

All Implement-   java.io.Serializable
ed Interfaces:
                    A Framework exception.

An InvalidSyntaxException object indicates that a filter string
parameter has an invalid syntax and cannot be parsed.

See Filter for a description of the filter string syntax.

2.21.13.1   Constructors
public **InvalidSyntaxException**(java.lang.String msg,
    java.lang.String filter)

Creates an exception of type InvalidSyntaxException.

This method creates an InvalidSyntaxException object with the specified message and the filter string which generated the exception.

Parameters: msg - The message.

filter - The invalid filter string.

2.21.13.2   **Methods**
public java.lang.String **getFilter**()

Returns the filter string that generated the InvalidSyntaxException object.

Returns: The invalid filter string.

See Also: getServiceReferences(String, String), addServiceListener(Service-Listener, String)

2.21.14   PackagePermission

public final class PackagePermission extends
java.security.BasicPermission

All Implement-   java.security.Guard, java.io.Serializable
ed Interfaces:
A bundle's authority to import or export a package.

A package is a dot-separated string that defines a fully qualified Java package.

For example:

    code>org.osgi.service.http

PackagePermission has two actions: EXPORT and IMPORT. The EXPORT action implies the IMPORT action.

2.21.14.1   **Fields**
public static final java.lang.String **EXPORT**

The action string EXPORT.

public static final java.lang.String **IMPORT**

The action string IMPORT.

2.21.14.2   **Constructors**
public **PackagePermission**(java.lang.String name, java.lang.String actions)

Defines the authority to import and/or export a package within the OSGi environment.

The name is specified as a normal Java package name: a dot-separated string. Wildcards may be used. For example:

```
org.osgi.service.http
javax.servlet.*
   *
```

Package Permissions are granted over all possible versions of a package. A bundle that needs to export a package must have the appropriate PackagePermission for that package; similarly, a bundle that needs to import a package must have the appropriate Package-Permssion for that package.

Permission is granted for both classes and resources.

Parameters: name - Package name.

                    actions - EXPORT, IMPORT (canonical order).

2.21.14.3   **Methods**

                    public boolean **equals**(java.lang.Object obj)

                    Determines the equality of two PackagePermission objects. This method checks that specified package has the same package name and PackagePermission actions as this PackagePermission object.

Overrides: java.security.BasicPermission.equals(java.lang.Object) in class java.security.BasicPermission

Parameters: obj - The object to test for equality with this PackagePermission object.

Returns: true if obj is a PackagePermission, and has the same package name and actions as this PackagePermission object; false otherwise.

                    public java.lang.String **getActions**()

                    Returns the canonical string representation of the Package-Permission actions.

                    Always returns present PackagePermission actions in the following order: EXPORT, IMPORT.

Overrides: java.security.BasicPermission.getActions() in class java.security.BasicPermission

Returns: Canonical string representation of the PackagePermission actions.

                    public int **hashCode**()

                    Returns the hash code value for this object.

Overrides:	java.security.BasicPermission.hashCode() in class java.security.BasicPermission
Returns:	A hash code value for this object.

public boolean **implies**(java.security.Permission p)

Determines if the specified permission is implied by this object.

This method checks that the package name of the target is implied by the package name of this object. The list of Package Permission actions must either match or allow for the list of the target object to imply the target Package Permission action.

The permission to export a package implies the permission to import the named package.

```
x.y.*, "export" -> x.y.z, "export"  is true
*, "import" -> x.y,  "import"        is true
*, "export" -> x.y,  "import"        is true
x.y, "export" -> x.y.z, "export"     is false
```

Overrides:	java.security.BasicPermission.implies(java.security.Permission) in class java.security.BasicPermission
Parameters:	p - The target permission to interrogate.
Returns:	true if the specified Package Permission action is implied by this object; false otherwise.

public java.security.PermissionCollection
    newPermissionCollection()

Returns a new Permission Collection object suitable for storing Package Permission objects.

Overrides:	java.security.BasicPermission.newPermissionCollection() in class java.security.BasicPermission
Returns:	A new Permission Collection object.

2.21.15 ServiceEvent

public class ServiceEvent extends java.util.EventObject

All Implemented Interfaces:	java.io.Serializable

A service lifecycle change event.

Service Event objects are delivered to a Service Listener objects when a change occurs in this service's lifecycle. A type code is used to identify the event type for future extendability.

OSGi reserves the right to extend the set of types.

See Also: ServiceListener

2.21.15.1   **Fields**

public static final int **MODIFIED**

The properties of a registered service have been modified.

This event is synchronously delivered *after* the service properties have been modified.

The value of MODIFIED is 0x00000002.

See Also: setProperties(Dictionary)

public static final int **REGISTERED**

This service has been registered.

This event is synchronously delivered *after* the service has been registered with the Framework.

The value of REGISTERED is 0x00000001.

See Also: registerService(String[], Object, Dictionary)

public static final int **UNREGISTERING**

This service is in the process of being unregistered.

This event is synchronously delivered *before* the service has completed unregistering.

If a bundle is using a service that is UNREGISTERING, the bundle should release its use of the service when it receives this event. If the bundle does not release its use of the service when it receives this event, the Framework will automatically release the bundle's use of the service while completing the service unregistration operation.

The value of UNREGISTERING is 0x00000004.

See Also: unregister(), ungetService(ServiceReference)

2.21.15.2   **Constructors**

public **ServiceEvent**(int type, ServiceReference reference)

Creates a new service event object.

Parameters: type - The event type.

reference - A ServiceReference object to the service that had a lifecycle change.

2.21.15.3   **Methods**

public ServiceReference **getServiceReference**()

Returns a reference to the service that had a change occur in its life-cycle.

This reference is the source of the event.

Returns: Reference to the service that had a lifecycle change.

public int **getType**()

Returns the type of event. The event type values are:

- REGISTERED
- MODIFIED
- UNREGISTERING

Returns: Type of service lifecycle change.

2.21.16   **ServiceFactory**

public interface ServiceFactory

Allows services to provide customized service objects in the OSGi environment.

When registering a service, a ServiceFactory object can be used instead of a service object, so that the bundle developer can gain control of the specific service object granted to a bundle that is using the service.

When this happens, the Bundle Context.get Service(Service-Reference) method calls the ServiceFactory.get Service method to create a service object specifically for the requesting bundle. The service object returned by the ServiceFactory object is cached by the Framework until the bundle releases its use of the service.

When the bundle's use count for the service equals zero (including the bundle stopping or the service being unregistered), the Service-Factory.unget Service method is called.

ServiceFactory objects are only used by the Framework and are not made available to other bundles in the OSGi environment.

See Also: getService(ServiceReference)

2.21.16.1   **Methods**

public java.lang.Object **getService**(Bundle bundle,
    ServiceRegistration registration)

Creates a new service object.

The Framework invokes this method the first time the specified bundle requests a service object using the BundleContext.get-Service(ServiceReference) method. The service factory can then return a specific service object for each bundle.

The Framework caches the value returned (unless it is null), and will return the same service object on any future call to Bundle-Context.getService from the same bundle.

The Framework will check if the returned service object is an instance of all the classes named when the service was registered. If not, then null is returned to the bundle.

Parameters: bundle - The bundle using the service.

registration - The ServiceRegistration object for the service.

Returns: A service object that *must* be an instance of all the classes named when the service was registered.

See Also: getService(ServiceReference)

public void **ungetService**(Bundle bundle, ServiceRegistration registration, java.lang.Object service)

Releases a service object.

The Framework invokes this method when a service has been released by a bundle. The service object may then be destroyed.

Parameters: bundle - The bundle releasing the service.

registration - The ServiceRegistration object for the service.

service - The service object returned by a previous call to the ServiceFactory.getService method.

See Also: ungetService(ServiceReference)

2.21.17   ServiceListener

public interface ServiceListener extends java.util.EventListener

All Superinter-faces: java.util.EventListener

A ServiceEvent listener.

ServiceListener is a listener interface that may be implemented by a bundle developer.

A ServiceListener object is registered with the Framework using the BundleContext.addServiceListener method. ServiceListener objects are called with a ServiceEvent object when a service has been registered or modified, or is in the process of unregistering.

Service Event object delivery to Service Listener objects is filtered by the filter specified when the listener was registered. If the Java Runtime Environment supports permissions, then additional filtering is done. Service Event objects are only delivered to the listener if the bundle which defines the listener object's class has the appropriate Service Permission to get the service using at least one of the named classes the service was registered under.

See Also: ServiceEvent, ServicePermission

2.21.17.1 Methods

public void **serviceChanged**(ServiceEvent event)

Receives notification that a service has had a lifecycle change.

Parameters: event - The Service Event object.

2.21.18 ServicePermission

**public final class ServicePermission extends
java.security.BasicPermission**

All Implemented Interfaces: java.security.Guard, java.io.Serializable

Indicates a bundle's authority to register or get a service.

- The REGISTER Service Permission action allows a bundle to register a service on the specified names.
- The GET Service Permission action allows a bundle to detect a service and get it.

Permission to get a service is required in order to detect events regarding the service. Untrusted bundles should not be able to detect the presence of certain services unless they have the appropriate Service Permission to get the specific service.

2.21.18.1 Fields

public static final java.lang.String **GET**

The action string get (Value is "get").

public static final java.lang.String **REGISTER**

The action string register (Value is "register").

2.21.18.2 Constructors

public **ServicePermission**(java.lang.String name, java.lang.String actions)

Create a new ServicePermission.

The name of the service is specified as a fully qualified class name.

```
ClassName ::= <class name> | <class name ending in
".*">
```

Examples:

```
org.osgi.service.http.HttpService
org.osgi.service.http.*
org.osgi.service.snmp.*
```

There are two possible actions: get and register. The get permission allows the owner of this permission to obtain a service with this name. The register permission allows the bundle to register a service under that name.

Parameters: name - class name

actions - get, register (canonical order)

2.21.18.3   **Methods**

public boolean **equals**(java.lang.Object obj)

Determines the equalty of two ServicePermission objects. Checks that specified object has the same class name and action as this ServicePermission.

Overrides: java.security.BasicPermission.equals(java.lang.Object) in class java.security.BasicPermission

Parameters: obj - The object to test for equality.

Returns: true if obj is a ServicePermission, and has the same class name and actions as this ServicePermission object; false otherwise.

public java.lang.String **getActions**()

Returns the canonical string representation of the actions. Always returns present actions in the following order: get, register.

Overrides: java.security.BasicPermission.getActions() in class java.security.BasicPermission

Returns: The canonical string representation of the actions.

public int **hashCode**()

Returns the hash code value for this object.

Overrides: java.security.BasicPermission.hashCode() in class java.security.BasicPermission

Returns: Hash code value for this object.

public boolean **implies**(java.security.Permission p)

Determines if a ServicePermission object "implies" the specified permission.

Overrides: java.security.BasicPermission.implies(java.security.Permission) in class java.security.BasicPermission

Parameters: p - The target permission to check.

Returns: true if the specified permission is implied by this object; false otherwise.

public java.security.PermissionCollection
   newPermissionCollection()

Returns a new PermissionCollection object for storing Service-Permission objects.

Overrides: java.security.BasicPermission.newPermissionCollection() in class java.security.BasicPermission

Returns: A new PermissionCollection object suitable for storing Service-Permission objects.

2.21.19   ServiceReference

public interface ServiceReference

A reference to a service.

The Framework returns ServiceReference objects from the Bundle-Context.getServiceReference and BundleContext.getService-References methods.

A ServiceReference may be shared between bundles and can be used to examine the properties of the service and to get the service object.

Every service registered in the Framework has a unique Service-Registration object and may have multiple, distinct Service-Reference objects referring to it. ServiceReference objects associated with a ServiceRegistration object have the same hash-Code and are considered equal (more specifically, their equals() method will return true when compared).

If the same service object is registered multiple times, Service-Reference objects associated with different ServiceRegistration objects are not equal.

See Also: getServiceReference(String), getServiceReferences(String, String), getService(ServiceReference)

2.21.19.1 **Methods**

public Bundle **getBundle**()

Returns the bundle that registered the service referenced by this ServiceReference object.

This method will always return null when the service has been unregistered. This can be used to determine if the service has been unregistered.

Returns: The bundle that registered the service referenced by this Service-Reference object; null if that service has already been unregistered.

See Also: registerService(String[], Object, Dictionary)

public java.lang.Object **getProperty**(java.lang.String key)

Returns the property value to which the specified property key is mapped in the properties Dictionary object of the service referenced by this ServiceReference object.

Property keys are case-insensitive.

This method will continue to return property values after the service has been unregistered. This is so references to unregistered services (for example, ServiceReference objects stored in the log) can still be interrogated.

Parameters: key - The property key.

Returns: The property value to which the key is mapped; null if there is no property named after the key.

public java.lang.String[] **getPropertyKeys**()

Returns an array of the keys in the properties Dictionary object of the service referenced by this ServiceReference object.

This method will continue to return the keys after the service has been unregistered. This is so references to unregistered services (for example, ServiceReference objects stored in the log) can still be interrogated.

This method is *case-preserving*; this means that every key in the returned array must have the same case as the corresponding key in the properties Dictionary that was passed to the registerService(String[], Object, Dictionary) or setProperties(Dictionary) methods.

Returns: An array of property keys.

public Bundle[] **getUsingBundles**()

Returns the bundles that are using the service referenced by this
Service Reference object. Specifically, this method returns the bundles whose usage count for that service is greater than zero.

Returns: An array of bundles whose usage count for the service referenced by
this Service Reference object is greater than zero; null if no bundles
are currently using that service.

Since: 1.1

2.21.20   ServiceRegistration

public interface ServiceRegistration

A registered service.

The Framework returns a Service Registration object when a
Bundle Context.register Service method is successful. The Service-
Registration object is for the private use of the registering bundle
and should not be shared with other bundles.

The Service Registration object may be used to update the properties of the service or to unregister the service.

See Also: registerService(String[], Object, Dictionary)

2.21.20.1   Methods
public ServiceReference **getReference**()
    throws java.lang.IllegalStateException

Returns a Service Reference object for a service being registered.

The Service Reference object may be shared with other bundles.

Returns: Service Reference object.

Throws: java.lang.IllegalStateException - If this Service Registration object
has already been unregistered.

public void **setProperties**(java.util.Dictionary properties)
    throws IllegalStateException, IllegalArgumentException

Updates the properties associated with a service.

The OBJECTCLASS and SERVICE_ID keys cannot be modified by this
method. These values are set by the Framework when the service is
registered in the OSGi environment.

The following steps are required to modify service properties:

1. The service's properties are replaced with the provided properties.

2. A service event of type MODIFIED is synchronously sent.

Parameters: properties - The properties for this service. See Constants for a list of standard service property keys. Changes should not be made to this object after calling this method. To update the service's properties this method should be called again.

Throws: IllegalStateException - If this ServiceRegistration object has already been unregistered.

IllegalArgumentException - If properties contains case variants of the same key name.

public void **unregister**()
    throws java.lang.IllegalStateException

Unregisters a service. Remove a ServiceRegistration object from the Framework service registry. All ServiceReference objects associated with this ServiceRegistration object can no longer be used to interact with the service.

The following steps are required to unregister a service:

1. The service is removed from the Framework service registry so that it can no longer be used. ServiceReference objects for the service may no longer be used to get a service object for the service.

2. A service event of type UNREGISTERING is synchronously sent so that bundles using this service can release their use of it.

3. For each bundle whose use count for this service is greater than zero:

4. The bundle's use count for this service is set to zero.

5. If the service was registered with a ServiceFactory object, the ServiceFactory.ungetService method is called to release the service object for the bundle.

Throws: java.lang.IllegalStateException - If this ServiceRegistration object has already been unregistered.

See Also: ungetService(ServiceReference), ungetService(Bundle, Service-Registration, Object)

2.21.21   **SynchronousBundleListener**

public interface SynchronousBundleListener extends
        BundleListener

All Superinter-   BundleListener, java.util.EventListener
faces:
                    A synchronous BundleEvent listener.

                    SynchronousBundleListener is a listener interface that may be
                    implemented by a bundle developer.

                    A SynchronousBundleListener object is registered with the Frame-
                    work using the addBundleListener(BundleListener) method.
                    SynchronousBundleListener objects are called with a BundleEvent
                    object when a bundle has been installed, started, stopped, updated,
                    or uninstalled.

                    Unlike normal BundleListener objects, SynchronousBundle-
                    Listeners are synchronously called during bundle life cycle process-
                    ing. The bundle life cycle processing will not proceed until all
                    SynchronousBundleListeners have completed. Synchronous-
                    BundleListener objects will be called prior to BundleListener
                    objects.

                    AdminPermission is required to add or remove a Synchronous-
                    BundleListener object.

Since: 1.1

See Also: BundleEvent

2.22   References

[3]   *The Standard for the Format of ARPA Internet Text Messages*
       STD 11, RFC 822, UDEL, August 1982
       ftp://ftp.isi.edu/in-notes/rfc822.txt

[4]   *The Hypertext Transfer Protocol - HTTP/1.1*
       RFC 2068 DEC, MIT/LCS, UC Irvine, January 1997
       ftp://ftp.isi.edu/in-notes/rfc2068.txt

[5]   *The Java 2 Platform API Specification*
       Standard Edition, Version 1.3, Sun Microsystems
       http://java.sun.com/j2se/1.3

[6]   *The Java Language Specification*
       Second Edition, Sun Microsystems, 2000
       http://java.sun.com/docs/books/jls/index.html

[7]   *A String Representation of LDAP Search Filters*
       RFC 1960, UMich, 1996
       ftp://ftp.isi.edu/in-notes/rfc1960.txt

[8]   *The Java Security Architecture for JDK 1.2*
Version 1.0, Sun Microsystems, October 1998
http://java.sun.com/products/jdk/1.2/docs/guide/security/spec/secu-
rity-spec.doc.html

[9]   *The Java 2 Package Versioning Specification*
http://java.sun.com/j2se/1.3/docs/guide/versioning/index.html

[10] *Codes for the Representation of Names of Languages*
ISO 639, International Standards Organization
http://lcweb.loc.gov/standards/iso639-2/iso639jac.html

[11] *Manifest Format*
http://java.sun.com/j2se/1.3/docs/guide/jar/
jar.html#JAR%20Manifest

[12] *RFC 822 Standard for the Format of ARPA Internet Text Messages*
Crocker, D., STD 11, RFC 822, UDEL, August 1982. http://
www.ietf.org/rfc/rfc0822.txt

3 Package Admin Service Specification

Version 1.0

3.1 Introduction

Bundles can export packages to other bundles. This creates a dependency between the bundle exporting a package and the bundle using this package. When the exporting bundle is uninstalled or updated, a decision must be taken regarding this shared package.

In the first version of the Framework, this decision was left to the implementation. Some implementations choose to do eager updates; Whenever a bundle was updated or uninstalled, its exported package were withdrawn and dependent bundles were stopped and resolved again. Other implementations choose to use a lazy update policy: leave the packages as they were and allow them to be used by other bundles until the Framework is restarted.

Leaving such an important detail to implementations was deemed unacceptable for this release. Therefore the Package Admin service is introduced. This service allows a management bundle to provide the policies for package sharing.

3.1.1 Essentials

- *Information* – Provide the status of all packages related to their sharing. This should include the importing bundles and exporting bundle.
- *Policy* – Allow a management bundle to provide a policy for package sharing when bundles are updated and uninstalled.
- *Minimal update* – Only bundles that depend of the package that needs to be resolved again will be restarted.

3.1.2 Operation

The Framework's system bundle can provide a Package Admin service for management bundles. The service is registered under the org.osgi.
service.packageadmin.PackageAdmin interface by the system bundle. It provides access to the internal structures of the Framework related to package sharing. See *Sharing Packages* on page 20.

The Framework always leaves the package sharing intact for packages exported by a bundle that is uninstalled or updated. Management bundles can choose to force the framework to eagerly update these packages using the Package Admin service. A policy of always using the most current packages exported by installed bundles can be implemented with a management bundle by watching Framework events for bundles being uninstalled or updated and refreshing the packages of those bundles using the Package Admin service.

3.1.3   Entities

- `PackageAdmin` – The interface that provides access to the internal Framework package sharing mechanism.

- `ExportedPackage` – Provides package information and its sharing status.
- *Management bundle* – A bundle that is provided by the operator to implement an operator specific policy.

Figure 11   *Class Diagram org.osgi.service.packageadmin*

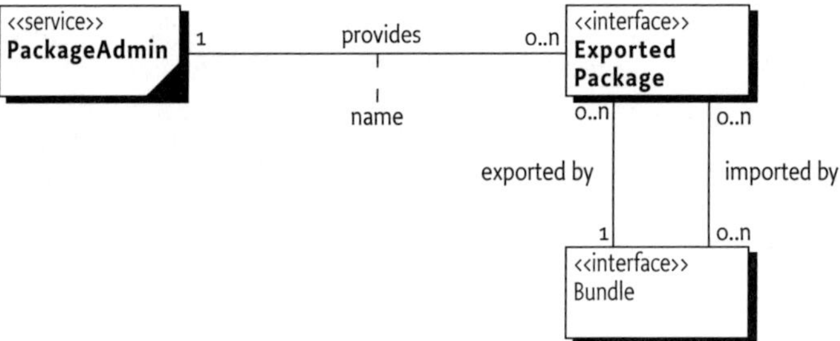

3.2   Package Admin

The Package Admin service is intended to allow a management bundle to define the policy for managing package sharing. It therefore provides methods for examining the status of the shared packages. It also allows the management bundle to refresh the packages, stopping and restarting bundles as necessary.

The `PackageAdmin` class provides the following methods:

- `getExportedPackage(String)` – Returns an `ExportedPackage` object that provides information about the requested package. This information can be used to make the decision to refresh the package.
- `getExportedPackages(Bundle)` – Returns a list of ExportedPackage objects for each package that the given bundle exports.

- `refreshPackages(Bundle[])` – The management bundle may call this method to refresh the exported packages of the specified bundles.

Information about the shared packages is provided by the Exported-Package objects. These objects provide detailed information about the bundles that import and export the package. This information can be used by a management bundle to guide its decisions.

3.3 Security

The Package Admin service is a *system service* that can easily be abused because it provides access to the internal data structures of the Framework. Only trusted bundles should have the ServicePermission[GET] for the PackageAdmin interface. No bundle must have ServicePermission[REGISTER], because only the Framework itself should register a system service.

This service should only be used by management bundles.

3.4 org.osgi.service.packageadmin

The OSGi Package Admin service Package. Specification Version 1.0.

Bundles wishing to use this package must list the package in the Import-Package header of the bundle's manifest. For example:

```
Import-Package: org.osgi.service.packageadmin;
specification-version=1.0
```

Class Summary

Interfaces

ExportedPackage	An exported package.
PackageAdmin	Framework service which allows bundle programmers to inspect the packages exported in the Framework and eagerly update or uninstall bundles.

3.4.1 ExportedPackage

public interface ExportedPackage

An exported package. Instances implementing this interface are created by the Package Admin service.

The information about an exported package provided by this object is valid only until the next time PackageAdmin.refreshPackages() is called. If an ExportedPackage object becomes stale (that is, the package it references has been updated or removed as a result of calling PackageAdmin.refreshPackages()), its getName() and getSpecificationVersion() continue to return their old values, isRemovalPending() returns true, and getExportingBundle() and getImportingBundles() return null.

3.4.1.1 Methods

public Bundle **getExportingBundle**()

Returns the bundle exporting the package associated with this ExportedPackage object.

Returns: The exporting bundle, or null if this ExportedPackage object has become stale.

public Bundle[] **getImportingBundles**()

Returns the resolved bundles that are currently importing the package associated with this ExportedPackage object.

The returned array always includes the bundle returned by getExportingBundle() since an exporter always implicitly imports its exported packages.

Returns: The array of resolved bundles currently importing the package associated with this ExportedPackage object, or null if this ExportedPackage object has become stale.

public java.lang.String **getName**()

Returns the name of the package associated with this ExportedPackage object.

Returns: The name of this ExportedPackage object.

public java.lang.String **getSpecificationVersion**()

Returns the specification version of this ExportedPackage, as specified in the exporting bundle's manifest file.

Returns: The specification version of this ExportedPackage object, or null if no version information is available.

public boolean **isRemovalPending**()

Returns true if the package associated with this ExportedPackage object has been exported by a bundle that has been updated or uninstalled.

Returns: true if the associated package is being exported by a bundle that has been updated or uninstalled, or if this ExportedPackage object has become stale; false otherwise.

3.4.2   PackageAdmin

public interface PackageAdmin

Framework service which allows bundle programmers to inspect the packages exported in the Framework and eagerly update or uninstall bundles. If present, there will only be a single instance of this service registered with the Framework.

The term *exported package* (and the corresponding interface ExportedPackage)refers to a package that has actually been exported (as opposed to one that is available for export).

The information about exported packages returned by this service is valid only until the next time refreshPackages(Bundle[]) is called. If an ExportedPackage object becomes stale, (that is, the package it references has been updated or removed as a result of calling Package-Admin.refreshPackages()), its getName() and getSpecificationVersion() continue to return their old values, isRemovalPending() returns true, and getExportingBundle() and getImportingBundles() return null.

3.4.2.1   Methods

public ExportedPackage **getExportedPackage**(java.lang.String
    name)

Gets the ExportedPackage object with the specified package name. All exported packages will be checked for the specified name. In an environment where the exhaustive list of packages on the system classpath is not known in advance, this method attempts to see if the named package is on the system classpath. This means that this method may discover an ExportedPackage object that was not present in the list returned by a prior call to getExportedPackages().

Parameters: name - The name of the exported package to be returned.

Returns: The exported package with the specified name, or null if no expored package with that name exists.

public ExportedPackage[] **getExportedPackages**(Bundle bundle)

Gets the packages exported by the specified bundle.

Parameters:　　bundle - The bundle whose exported packages are to be returned, or null if all the packages currently exported in the Framework are to be returned. If the specified bundle is the system bundle (that is, the bundle with id 0), this method returns all the packages on the system classpath whose name does not start with "java.". In an environment where the exhaustive list of packages on the system classpath is not known in advance, this method will return all currently known packages on the system classpath, that is, all packages on the system classpath that contains one or more classes that have been loaded.

Returns:　　The array of packages exported by the specified bundle, or null if the specified bundle has not exported any packages.

public void **refreshPackages**(Bundle[] bundles)
　　throws SecurityException

Forces the update (replacement) or removal of packages exported by the specified bundles.

If no bundles are specified, this method will update or remove any packages exported by any bundles that were previously updated or uninstalled. The technique by which this is accomplished may vary among different Framework implementations. One permissible implementation is to stop and restart the Framework.

This method returns to the caller immediately and then performs the following steps in its own thread:

1. Compute a graph of bundles starting with the specified ones. If no bundles are specified, compute a graph of bundles starting with previously updated or uninstalled ones. Any bundle that imports a package that is currently exported by a bundle in the graph is added to the graph. The graph is fully constructed when there is no bundle outside the graph that imports a package from a bundle in the graph. The graph may contain UNINSTALLED bundles that are currently still exporting packages.

2. Each bundle in the graph will be stopped as described in the Bundle.stop method.

3. Each bundle in the graph that is in the RESOLVED state is moved to the INSTALLED state. The effect of this step is that bundles in the graph are no longer RESOLVED.

4. Each bundle in the graph that is in the UNINSTALLED state is removed from the graph and is now completely removed from the Framework.

5. Each bundle in the graph that was in the ACTIVE state prior to Step 2 is started as described in the Bundle.start method, causing all bundles required for the restart to be resolved. It is possible that, as a result of the previous steps, packages that were previously exported no longer are. Therefore, some bundles may be unresolvable until another bundle offering a compatible package for export has been installed in the Framework.

For any exceptions that are thrown during any of these steps, a FrameworkEvent of type ERROR is broadcast, containing the exception.

Parameters: bundles - the bundles whose exported packages are to be updated or removed, or null for all previously updated or uninstalled bundles.

Throws: SecurityException - if the caller does not have the AdminPermission and the Java runtime environment supports permissions.

4 Permission Admin Service Specification

Version 1.0

4.1 Introduction

In the Framework, a bundle can have a single set of permissions associated with it. These permissions are used to verify that a bundle is authorized to execute privileged code. For example, a FilePermission defines what files can be used and in what way.

The policy of providing the permissions to the bundle should be delegated to a management bundle. The Framework provides the Permission Admin service so that a management bundle can administrate bundle's permissions and provide defaults for all bundles.

The related mechanisms of the Framework are discussed in *Security* on page 52.

4.1.1 Essentials

- *Status information* – Provide status information about the permissions of a bundle.
- *Administrate* – Allow a management bundle to set the permissions before, during or after it is installed.
- *Defaults* – Provide access to the default permissions. These are the permissions when a bundle has no specific permissions set.

4.1.2 Operation

The Framework maintains a repository of permissions. These permissions are stored under the bundle location string. Using the bundle location allows for setting the permissions *before* a bundle is downloaded. The Framework must consult this repository when it needs the permissions of a bundle. When no specific permissions are set, the bundle must use the default permissions. If no default is set, it must use java.security.AllPermission.

The Permission Admin service is registered by the Framework's system bundle under the org.osgi.service.permissionadmin.PermissionAdmin interface. This is an optional singleton service, so at most one Permission Admin service is registered at any moment in time.

The Permission Admin service provides access to the permission repository. A management bundle can get, set, update and deleted permissions from this repository. A management bundle can use the SynchronousBundleListener to set the permissions during the installation or updating of a bundle.

4.1.3 Entities

- PermissionAdmin – The service that provides access to the permission repository of the Framework.
- PermissionInfo – An object that holds the information to construct a Permission object.
- *Bundle location* – The string that specifies the bundle location, this is described in *Bundle Location* on page 29.
- *Management bundle* – A bundle that is provided by the operator to implement an operator specific policy.

Figure 12   *Class Diagram org.osgi.service.permissionadmin.*

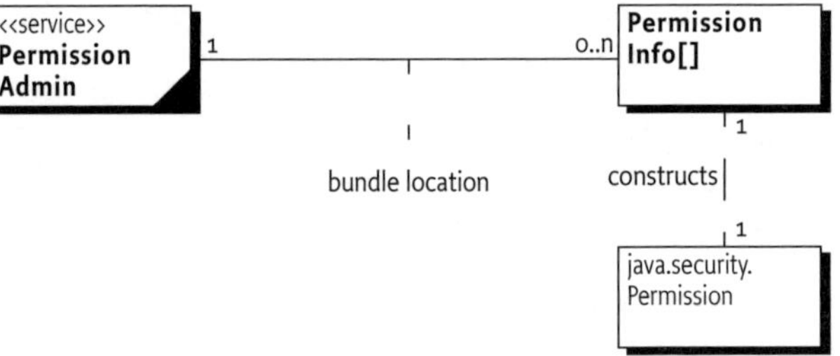

4.2 Permission Admin service

The Permission Admin service needs to manipulate the default permissions and the permissions associated with a specific bundle. The default permissions and the bundle-specific permissions are stored persistently. It is possible to set a bundle's permissions before the bundle is installed in the Framework because the bundle's location is used to set the bundle's permissions.

The manipulation of a bundle's permission, however, may also be done in real time, when a bundle is downloaded, or set just before the bundle is downloaded. To support this flexibility, a SynchronousBundleListener object may be used to allow a management bundle to detect the installation or update of a new bundle, and set the required permissions before the installation completes.

Permissions are activated the first time a permission check for a bundle is performed, meaning that if a bundle has opened a file, this file must remain usable even if later in time the permission to open that file is removed.

Permission information is *not* specified using java.security.Permission objects. The reason for this approach is the relationship between the required persistence of the information across Framework restarts and the concept of classloaders in the Framework.

Permission objects must be subclasses of Permission and may be exported from any bundle. The Framework can access these permissions as long as they are exported, but the management bundle would have to import all possible packages that contain permissions. This requirement would limit the permission types. Therefore, the Permission Admin service uses the PermissionInfo class to specify permission information. Objects of this class are used to create Permission objects.

PermissionInfo objects restrict the possible Permission objects that can be used. A Permission class can only be described by a PermissionInfo object when it fulfills the following characteristics:

- It must be a subclass of java.security.Permission.
- It must use the two-argument public constructor type(classname,actions).
- The type name of the permission must be a valid, fully qualified class name.
- The class for this type name must be available to the Framework code from the system classpath or from an exported package, so it can be loaded by the Framework.
- The class must be public.

The permissions are always set as an array of PermissionInfo objects to make the assignment atomic.

The PermissionAdmin interface provides the following methods:

- getLocations() – Returns a list of locations that have permissions assigned to them. This method allows a management bundle to examine the current permissions.
- getPermissions(String) – Returns a list of PermissionInfo objects that are set for that location or null if no permissions are set.
- setPermissions(String, PermissionInfo[]) – Associates permissions with a specific location. or null when the permissions should be removed.
- getDefaultPermissions() – This method returns the set of default permissions.

- setDefaultPermissions(PermissionInfo[]) – This method sets the default permissions.

4.3 Security

The Permission Admin is a system service and is very sensitive to abuse. A bundle that can access and use the Permission Admin service has full control over the OSGi environment. Only bundles that are fully trusted should have ServicePermission[GET] for this service.

No bundle must have ServicePermission[REGISTER] for this service because only the Framework should provide this service.

4.4 org.osgi.service.permissionadmin

The OSGi Permission Admin service Package. Specification Version 1.0.

Bundles wishing to use this package must list the package in the Import-Package header of the bundle's manifest. For example:

```
Import-Package: org.osgi.service.permissionadmin;
specification-version=1.0
```

Class Summary

Interfaces

PermissionAdmin	The Permission Admin service allows operators to manage the permissions of bundles.

Classes

PermissionInfo	Permission representation used by the Permission Admin service.

4.4.1 PermissionAdmin

public interface PermissionAdmin

The Permission Admin service allows operators to manage the permissions of bundles. There is at most one Permission Admin service present in the Framework.

Access to the Permission Admin service is protected by corresponding ServicePermission. In addition the AdminPermission is required to actually set permissions.

Bundle permissions are managed using a permission table. A bundle's location serves as the key into this permission table. The value of a table entry is the set of permissions (of type PermissionInfo) granted to the bundle with the given location. A bundle may have an entry in the permission table prior to being installed in the Framework.

The permissions specified in setDefaultPermissions are used as the default permissions which are granted to all bundles that do not have an entry in the permission table.

Any changes to a bundle's permissions in the permission table will take effect no later than when bundle's java.security.Protection-Domain is involved in a permission check, and will be made persistent.

Only permission classes on the system classpath or from an exported package are considered during a permission check. Additionally, only permission classes that are subclasses of java.security.Permission and define a 2-argument constructor that takes a *name* string and an *actions* string can be used.

Permissions implicitly granted by the Framework (for example, a bundle's permission to access its persistent storage area) cannot be changed, and are not reflected in the permissions returned by getPermissions and getDefaultPermissions.

4.4.1.1   **Methods**

public PermissionInfo[] **getDefaultPermissions**()

Gets the default permissions.

These are the permissions granted to any bundle that does not have permissions assigned to its location.

Returns: The default permissions, or null if default permissions have not been defined.

public java.lang.String[] **getLocations**()

Returns the bundle locations that have permissions assigned to them, that is, bundle locations for which an entry exists in the permission table.

Returns: The locations of bundles that have been assigned any permissions, or null if the permission table is empty.

public PermissionInfo[] **getPermissions**(java.lang.String location)

Gets the permissions assigned to the bundle with the specified location.

Parameters: location - The location of the bundle whose permissions are to be returned.

Returns: The permissions assigned to the bundle with the specified location, or null if that bundle has not been assigned any permissions.

public void **setDefaultPermissions**(PermissionInfo[] permissions)
    throws SecurityException

Sets the default permissions.

These are the permissions granted to any bundle that does not have permissions assigned to its location.

Parameters: permissions - The default permissions.

Throws: SecurityException - if the caller does not have the Admin Permission.

public void **setPermissions**(java.lang.String location,
    PermissionInfo[] perms)
    throws SecurityException

Assigns the specified permissions to the bundle with the specified location.

Parameters: location - The location of the bundle that will be assigned the permissions.

perms - The permissions to be assigned, or null if the specified location is to be removed from the permission table.

Throws: SecurityException - if the caller does not have the Admin Permission.

4.4.2   PermissionInfo

public class PermissionInfo

Permission representation used by the Permission Admin service.

This class encapsulates three pieces of information: a Permission *type* (class name), which must be a subclass of java.security.Permission, and the *name* and *actions* arguments passed to its constructor.

In order for a permission represented by a PermissionInfo to be instantiated and considered during a permission check, its Permission class must be available from the system classpath or an exported package. This means that the instantiation of a permission represented by a PermissionInfo may be delayed until its Permission class has been exported to the Framework.

4.4.2.1   **Constructors**

public **PermissionInfo**(java.lang.String encodedPermission)
  throws java.lang.IllegalArgumentException

Constructs a PermissionInfo object from the given encoded PermissionInfo string.

Parameters: encodedPermission - The encoded PermissionInfo.

Throws: java.lang.IllegalArgumentException - if encodedPermission is not properly formatted.

See Also: getEncoded()

public **PermissionInfo**(java.lang.String type, java.lang.String name,
  java.lang.String actions)

Constructs a PermissionInfo from the given type, name, and actions.

Parameters: type - The fully qualified class name of the permission represented by this PermissionInfo. The class must be a subclass of java.security.Permission and must define a 2-argument constructor that takes a *name* string and an *actions* string.

name - The permission name that will be passed as the first argument to the constructor of the Permission class identified by type.

actions - The permission actions that will be passed as the second argument to the constructor of the Permission class identified by type.

4.4.2.2   **Methods**

public boolean **equals**(java.lang.Object obj)

Determines the equality of two PermissionInfo objects. This method checks that specified object has the same type, name and actions as this PermissionInfo object.

Overrides: java.lang.Object.equals(java.lang.Object) in class java.lang.Object

Parameters: obj - The object to test for equality with this PermissionInfo object.

Returns: true if obj is a PermissionInfo, and has the same type, name and actions as this PermissionInfo object; false otherwise.

public final java.lang.String **getActions**()

Returns the actions of the permission represented by this PermissionInfo.

Returns: The actions of the permission represented by this PermissionInfo, or null if the permission does not have any actions associated with it.

public final java.lang.String **getEncoded**()

Returns the string encoding of this PermissionInfo in a form suitable for restoring this PermissionInfo.

The encoding format is:

    (type)

or

    (type "*name*")

or

    (type "*name*" "*actions*")

where *name* and *actions* are strings that are encoded for proper parsing. Specifically, the ", \, carriage return, and linefeed characters are escaped using \", \\, \r, and \n, respectively.

Returns: The string encoding of this PermissionInfo.

public final java.lang.String **getName**()

Returns the name of the permission represented by this PermissionInfo.

Returns: The name of the permission represented by this PermissionInfo, or null if the permission does not have a name.

public final java.lang.String **getType**()

Returns the fully qualified class name of the permission represented by this PermissionInfo.

Returns: The fully qualified class name of the permission represented by this PermissionInfo.

public int **hashCode**()

Returns the hash code value for this object.

Overrides: java.lang.Object.hashCode() in class java.lang.Object

Returns: A hash code value for this object.

public java.lang.String **toString**()

Returns the string representation of this Permission Info. The string is created by calling the get Encoded method on this Permission-Info.

Overrides: java.lang.Object.toString() in class java.lang.Object

Returns: The string representation of this Permission Info.

5   Service Tracker Specification

Version 1.1

5.1   Introduction

The Framework provides a powerful, but very dynamic, programming environment. Bundles are installed, started, stopped, updated and uninstalled without shutting down the Framework. Dependencies between bundles are monitored by the Framework, but bundles must cooperate in handling these dependencies correctly.

An important aspect of the Framework is the service registry. Bundle developers must be careful not to use service objects that have been unregistered. The dynamic nature of the Framework service registry makes it necessary to track the service objects as they are registered and unregistered. It is easy to overlook rare race conditions or boundary conditions that will lead to random errors.

An example of a non-trivial problem is to create the initial list of services of a certain type when a bundle is started. When the Service-Listener object is registered before the Framework is asked for the list of services, without special precautions, duplicates can enter the list. When the ServiceListener object is registered after the list is made, it is possible to miss relevant events.

The specification defines a utility class, ServiceTracker, that makes tracking the registration, modification, and unregistration of services much easier. A ServiceTracker class can be customized by implementing the ServiceTrackerCustomizer interface or by subclassing the ServiceTracker class.

This utility specifies a class that significantly reduces the complexity to track services in the service registry.

5.1.1   Essentials

- *Customizing* – Allow a default implementation to be customized so that bundle developers can easily start simple but extend it to their needs.
- *Small* – Every Framework implementation should have this utility implemented. It should therefore be very small because

some Framework implementations target minimal OSGi environments.

- *Tracked set* – Track a single object defined by a ServiceReference object, all instances of a service or any set specified by a filter expression.

5.1.2 Operation

The fundamental tasks of a ServiceTracker object are:

- To create an initial list of services as specified by its owner.
- To listen to ServiceEvent instances so that services of interest to the owner are properly tracked.
- To allow the owner to customize the tracking process by allowing programmatic selection of the services to be tracked as well as act when a service is added or removed.

A ServiceTracker object populates a set of services that match a given search criteria, and then listens to ServiceEvent objects which correspond to those services.

5.1.3 Entities

Figure 13   *Class diagram of org.osgi.util.tracker*

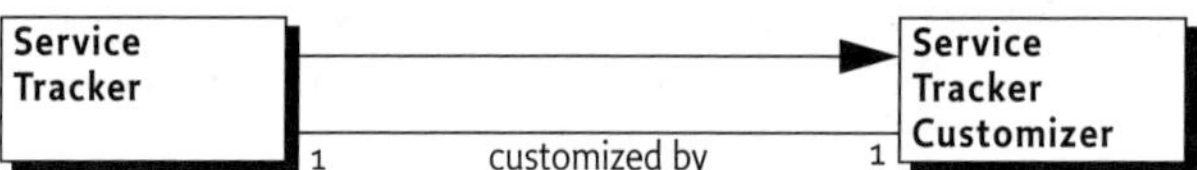

5.2 ServiceTracker Class

The ServiceTracker interface defines three constructors to create ServiceTracker objects, each providing different search criteria:

- ServiceTracker(BundleContext, String, ServiceTrackerCustomizer) – This constructor takes a service interface name as the search criterion. The ServiceTracker object must then track all services that are registered under the specified service interface name.
- ServiceTracker(BundleContext, Filter, ServiceTrackerCustomizer) – This constructor uses a Filter object to specify the services to be tracked. The ServiceTracker must then track all services that match the specified filter.
- ServiceTracker(BundleContext, ServiceReference, ServiceTrackerCustomizer) – This constructor takes a ServiceReference object as the search criterion. The ServiceTracker must then track only the service that corresponds to the specified ServiceReference. Using this constructor, no more than one service must

ever be tracked, because a ServiceReference refers to a specific service.

Each of the ServiceTracker constructors takes a BundleContext object as a parameter. This BundleContext object must be used by a ServiceTracker object to track, get, and unget services.

A new ServiceTracker object must not begin tracking services until its open method is called.

5.3 Using a Service Tracker

Once a ServiceTracker object is opened, it begins tracking services immediately. A number of methods are available to the bundle developer to monitor the services that are being tracked. The ServiceTracker class defines these methods:

- getService() – Returns one of the services being tracked or null if there are no active services being tracked.
- getServices() – Returns an array of all the tracked services. The number of tracked services is returned by the size method.
- getServiceReference() – Returns a ServiceReference object for one of the services being tracked. The service object for this service may be returned by calling the ServiceTracker object's getService() method.
- getServiceReferences() – Returns a list of the ServiceReference objects for services being tracked. The service object for a specific tracked service may be returned by calling the ServiceTracker object's getService(ServiceReference) method.
- waitForService(long) – Allows the caller to wait until at least one instance of a service is tracked or until the timeout expires. If the timeout is zero, the caller must wait until at least one instance of a service is tracked. waitForService must not used within the BundleActivator methods, as these methods are expected to complete in a short period of time. A Framework may wait until the start method returns before it starts the bundle that registers the service that is waited for, creating a dead-lock situation.
- remove(ServiceReference) – This method may be used to remove a specific service from being tracked by the ServiceTracker object, causing removedService to be called for that service.
- close() – This method must remove all services being tracked by the ServiceTracker object, causing removedService to be called for all tracked services.

5.4 Customizing the ServiceTracker class

The behavior of the ServiceTracker class can be customized by either providing a ServiceTrackerCustomizer object implementing the desired behavior when the ServiceTracker object is constructed, or by subclassing the ServiceTracker class and overriding the ServiceTrackerCustomizer methods.

The ServiceTrackerCustomizer interface defines these methods:

- addingService(ServiceReference) – Called whenever a service is being added to the ServiceTracker object.
- modifiedService(ServiceReference, Object) – Called whenever a tracked service is modified.
- removedService(ServiceReference, Object) – Called whenever a tracked service is removed from the ServiceTracker object.

When a service is being added to the ServiceTracker object or when a tracked service is modified or removed from the ServiceTracker object, it must call addingService, modifiedService, or removedService, respectively, on the ServiceTrackerCustomizer object (if specified when the ServiceTracker object was created); otherwise it must call these methods on itself.

A bundle developer may customize the action when a service is tracked. Another reason for customizing the ServiceTracker class is to programmatically select which services are tracked. A filter may not sufficiently specify the services that the bundle developer is interested in tracking. By implementing addingService, the bundle developer can use additional runtime information to determine if the service should be tracked. If null is returned by the addingService method, the service must not be tracked.

Finally, the bundle developer can return a specialized object from addingService that differs from the service object. This specialized object could contain the service object and any associated information. This returned object is then tracked instead of the service object. When the removedService method is called, the object that is passed along with the ServiceReference object is the one that was returned from the earlier call to the addingService method.

5.5 Customizing Example

An example of customizing the action taken when a service is tracked might be registering a Servlet with each Http Service that is tracked. This customization could be done by sub-classing the ServiceTracker class and overriding the addingService and removedService methods as follows:

```
public Object addingService( ServiceReference
reference) {
   Object obj = context.getService(reference);
   HttpService svc = (HttpService)obj;
   // Register the Servlet using svc
   ...
   return svc;
}
public void removedService( ServiceReference
reference,
   Object obj ){
   HttpService svc = (HttpService)obj;
   // Unregister the Servlet using svc
   ...
   context.ungetService(reference);
}
```

5.6 Security

A ServiceTracker object contains a BundleContext instance variable that is accessible to the methods in a subclass. A BundleContext object should never be given to other bundles because it is used for security aspects of the Framework.

The ServiceTracker implementation does not have a method to get the BundleContext object but subclasses should be careful not to provide such a method if the ServiceTracker object is given to other bundles.

The services that are being tracked are available via a ServiceTracker. These services are dependent on the BundleContext as well. It is therefore necessary to do a careful security analysis when ServiceTracker objects are given to other bundles.

5.7 org.osgi.util.tracker

The OSGi Service Tracker Package. Specification Version 1.1.

Bundles wishing to use this package must list the package in the Import-Package header of the bundle's manifest. For example:

```
Import-Package: org.osgi.util.tracker; specification-
version=1.1
```

Class Summary

Interfaces

ServiceTrackerCus- tomizer	The ServiceTrackerCustomizer interface allows a Service-Tracker object to customize the service objects that are tracked.

Classes

ServiceTracker	The ServiceTracker class simplifies using services from the Framework's service registry.

5.7.1 ServiceTracker

public class ServiceTracker implements
 ServiceTrackerCustomizer

All Implement-
ed Interfaces: ServiceTrackerCustomizer

The ServiceTracker class simplifies using services from the Framework's service registry.

A ServiceTracker object is constructed with search criteria and a ServiceTrackerCustomizer object. A ServiceTracker object can use the ServiceTrackerCustomizer object to customize the service objects to be tracked. The ServiceTracker object can then be opened to begin tracking all services in the Framework's service registry that match the specified search criteria. The ServiceTracker object correctly handles all of the details of listening to ServiceEvent objects and getting and ungetting services.

The getServiceReferences method can be called to get references to the services being tracked. The getService and getServices methods can be called to get the service objects for the tracked service.

5.7.1.1 Fields
protected final BundleContext **context**

Bundle context this ServiceTracker object is tracking against.

protected final Filter **filter**

Filter specifying search criteria for the services to track.

Since:　　　　1.1

5.7.1.2　　　**Constructors**

public **ServiceTracker**(BundleContext context, Filter filter,
　　ServiceTrackerCustomizer customizer)

Create a ServiceTracker object on the specified Filter object.

Services which match the specified Filter object will be tracked by
this ServiceTracker object.

Parameters:　　context - BundleContext object against which the tracking is done.

filter - Filter object to select the services to be tracked.

customizer - The customizer object to call when services are added,
modified, or removed in this ServiceTracker object. If customizer is
null, then this ServiceTracker object will be used as the Service-
TrackerCustomizer object and the ServiceTracker object will call
the ServiceTrackerCustomizer methods on itself.

Since:　　　　1.1

public **ServiceTracker**(BundleContext context, ServiceReference
　　reference, ServiceTrackerCustomizer customizer)

Create a ServiceTracker object on the specified ServiceReference
object.

The service referenced by the specified ServiceReference object will
be tracked by this ServiceTracker object.

Parameters:　　context - BundleContext object against which the tracking is done.

reference - ServiceReference object for the service to be tracked.

customizer - The customizer object to call when services are added,
modified, or removed in this ServiceTracker object. If customizer is
null, then this ServiceTracker object will be used as the Service-
TrackerCustomizer object and the ServiceTracker object will call
the ServiceTrackerCustomizer methods on itself.

public **ServiceTracker**(BundleContext context, java.lang.String
　　clazz, ServiceTrackerCustomizer customizer)

Create a ServiceTracker object on the specified class name.

Services registered under the specified class name will be tracked by
this ServiceTracker object.

Parameters: context - Bundle Context object against which the tracking is done.

clazz - Class name of the services to be tracked.

customizer - The customizer object to call when services are added, modified, or removed in this Service Tracker object. If customizer is null, then this Service Tracker object will be used as the Service-Tracker Customizer object and the Service Tracker object will call the Service Tracker Customizer methods on itself.

5.7.1.3   Methods

public java.lang.Object **addingService**(ServiceReference reference)

Default implementation of the Service Tracker Customizer.adding-Service method.

This method is only called when this Service Tracker object has been constructed with a null Service Tracker Customizer argument. The default implementation returns the result of calling get Service, on the Bundle Context object with which this Service Tracker object was created, passing the specified Service Reference object.

This method can be overridden to customize the service object to be tracked for the service being added.

Specified By: addingService(ServiceReference) in interface ServiceTracker

Parameters: reference - Reference to service being added to this Service Tracker object.

Returns: The service object to be tracked for the service added to this Service-Tracker object.

public synchronized void **close**()

Close this Service Tracker object.

This method should be called when this Service Tracker object should end the tracking of services.

protected void **finalize**()

Properly close this Service Tracker object when finalized. This method calls the close method to close this Service Tracker object if it has not already been closed.

Overrides: java.lang.Object.finalize() in class java.lang.Object

Throws: Throwable

public java.lang.Object **getService**()

Returns a service object for one of the services being tracked by this Service Tracker object.

If any services are being tracked, this method returns the result of calling get Service(get Service Reference()).

Returns:　Service object or null if no service is being tracked.

public java.lang.Object **getService**(ServiceReference reference)

Returns the service object for the specified Service Reference object if the referenced service is being tracked by this Service Tracker object.

Parameters:　reference - Reference to the desired service.

Returns:　Service object or null if the service referenced by the specified Service Reference object is not being tracked.

public ServiceReference **getServiceReference**()

Returns a Service Reference object for one of the services being tracked by this Service Tracker object.

If multiple services are being tracked, the service with the highest ranking (as specified in its service.ranking property) is returned.

If there is a tie in ranking, the service with the lowest service ID (as specified in its service.id property); that is, the service that was registered first is returned.

This is the same algorithm used by Bundle Context.get Service-Reference.

Returns:　Service Reference object or null if no service is being tracked.

Since:　1.1

public ServiceReference[] **getServiceReferences**()

Return an array of Service Reference objects for all services being tracked by this Service Tracker object.

Returns:　Array of Service Reference objects or null if no service are being tracked.

public java.lang.Object[] **getServices**()

Return an array of service objects for all services being tracked by this Service Tracker object.

Returns: Array of service objects or null if no service are being tracked.

public void **modifiedService**(ServiceReference reference,
    java.lang.Object service)

Default implementation of the Service Tracker-
Customizer.modified Service method.

This method is only called when this Service Tracker object has
been constructed with a null Service Tracker Customizer argument.
The default implementation does nothing.

Specified By: modifiedService(ServiceReference, Object) in interface Service-
Tracker

Parameters: reference - Reference to modified service.

service - The service object for the modified service.

public synchronized void **open**()
    throws java.lang.IllegalStateException

Open this Service Tracker object and begin tracking services.

Services which match the search criteria specified when this
Service Tracker object was created are now tracked by this Service-
Tracker object.

Throws: java.lang.IllegalStateException - if the Bundle Context object with
which this Service Tracker object was created is no longer valid.

public void **remove**(ServiceReference reference)

Remove a service from this Service Tracker object. The specified service will be removed from this Service Tracker object. If the specified service was being tracked then the
Service Tracker Customizer.removed Service method will be called
for that service.

Parameters: reference - Reference to the service to be removed.

public void **removedService**(ServiceReference reference,
    java.lang.Object object)

Default implementation of the Service Tracker-
Customizer.removed Service method.

This method is only called when this Service Tracker object has been constructed with a null Service Tracker Customizer argument. The default implementation calls unget Service, on the Bundle-Context object with which this Service Tracker object was created, passing the specified Service Reference object.

Specified By: removedService(ServiceReference, Object) in interface Service-Tracker

Parameters: reference - Reference to removed service.

service - The service object for the removed service.

public int **size**()

Return the number of services being tracked by this Service Tracker object.

Returns: Number of services being tracked.

public java.lang.Object **waitForService**(long timeout)

Wait for at least one service to be tracked by this Service Tracker object.

It is strongly recommended that wait For Service is not used during the calling of the Bundle Activator methods. Bundle Activator methods are expected to complete in a short period of time.

Parameters: timeout - time interval in milliseconds to wait. If zero, the method will wait indefinately.

Returns: Returns the result of get Service().

Throws: InterruptedException

5.7.2   ServiceTrackerCustomizer

public interface ServiceTrackerCustomizer

All Known Im-plementing Classes: ServiceTracker

The Service Tracker Customizer interface allows a Service Tracker object to customize the service objects that are tracked. The Service-Tracker Customizer object is called when a service is being added to the Service Tracker object. The Service Tracker Customizer can then return an object for the tracked service. The Service Tracker-Customizer object is also called when a tracked service is modified or has been removed from the Service Tracker object.

5.7.2.1

Methods

public java.lang.Object **addingService**(ServiceReference reference)

A service is being added to the Service Tracker object.

This method is called before a service which matched the search parameters of the Service Tracker object is added to it. This method should return the service object to be tracked for this Service-Reference object. The returned service object is stored in the Service Tracker object and is available from the get Service and get-Services methods.

Parameters: reference - Reference to service being added to the Service Tracker object.

Returns: The service object to be tracked for the Service Reference object or null if the Service Reference object should not be tracked.

public void **modifiedService**(ServiceReference reference,
    java.lang.Object service)

A service tracked by the Service Tracker object has been modified.

This method is called when a service being tracked by the Service-Tracker object has had it properties modified.

Parameters: reference - Reference to service that has been modified.

service - The service object for the modified service.

public void **removedService**(ServiceReference reference,
    java.lang.Object service)

A service tracked by the Service Tracker object has been removed.

This method is called after a service is no longer being tracked by the Service Tracker object.

Parameters: reference - Reference to service that has been removed.

service - The service object for the removed service.

6 Log Service Specification

Version 1.1

6.1 Introduction

The Log Service provides a general purpose message logger for the OSGi environment. It consists of two services, one for logging information and another for retrieving log information from the past or as they are recorded.

This specification defines the methods and semantics of interfaces which bundle developers can use to log entries and to get log entries.

Bundles can use the Log Service to log interesting information for the operator. Other bundles, oriented toward management of the environment, can use the Log Reader service to retrieve Log Entry objects that happened recently or to receive Log Entry objects as they are logged by other bundles.

6.1.1 Entities

- `LogService` – The service interface that allows a bundle to log information. The information can include a message, a level, an exception, a `ServiceReference` object, and a `Bundle` object.
- `LogEntry` - An interface that allows access to a log entry in the log. It includes all the information that can be logged through the Log Service and an additional time stamp.
- `LogReaderService` - A service interface that allows access to a list of recent `LogEntry` objects, and allows the registration of a `LogListener` object that receives `LogEntry` objects as they are created.
- `LogListener` - The interface for the listener to `LogEntry` objects. Must be registered with the Log Reader Service.

Figure 14   *Log Service Class Diagram org.osgi.service.log package*

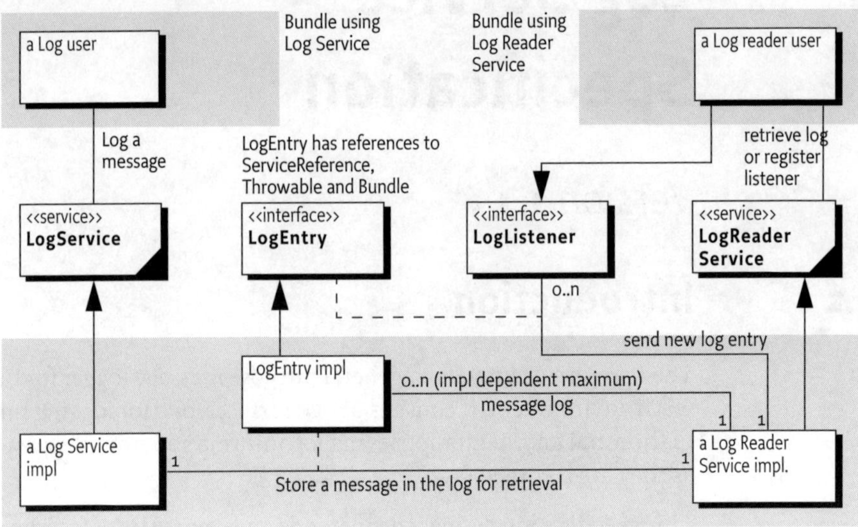

6.2   The Log Service Interface

The LogService interface allows bundle developers to log messages
that can be distributed to other bundles, which in turn can forward
the logged entries to a file system, remote systems, or some other
destination.

The LogService interface allows the bundle developer to:

- Specify a message and/or exception to be logged.
- Supply a log level representing the severity of the message being
  logged.
- Specify the Service associated with the log requests.

By obtaining a LogService object from the Framework service registry, a bundle can start logging messages to it by calling one of the
LogService methods. A Log Service object can log any message; it is
primarily intended, however, for reporting events and error conditions.

The LogService interface defines these methods for logging messages:

- log(int, String) – This method logs a simple message at a given
  log level.
- log(int, String, Throwable) – This method logs a message with
  an exception at a given log level.
- log(ServiceReference, int, String) – This method logs a message
  associated with a specific service.

- log(ServiceReference, int, String, Throwable) – This method logs a message with an exception associated with a specific service.

While it is possible for a bundle to call one of the log methods without providing a ServiceReference object, it is recommended that the caller supply the ServiceReference argument whenever appropriate, because it provides important context information to the operator in case of problems.

The following example demonstrates the use of a log method to write a message into the log.

```
logService.log(
    myServiceReference,
    LogService.LOG_INFO,
    "myService is up and running"
);
```

In the example, the myServiceReference parameter identifies the service associated with the log request. The specified level, LogService.LOG_INFO, indicates that this message is informational.

The following example code records error conditions as log messages.

```
try {
    FileInputStream fis = new FileInputStream(
"myFile");
    int b;
    while ( (b = fis.read()) != -1 ) {
        ...
    }
    fis.close();
}
catch ( IOException exception ) {
    logService.log(
        myServiceReference,
        LogService.LOG_ERROR,
        "Cannot access file",
        exception );
}
```

Notice that in addition to the error message, the exception itself is also logged. Providing this information can significantly simplify problem determination by the operator.

6.3   Log Level and Error Severity

The log methods expect a log level indicating error severity, which can be used to filter log messages when they are retrieved. The severity levels are defined in the LogService interface.

Callers must supply the log levels that they deem appropriate when making log requests. The following table lists the log levels.

Level	Descriptions
LOG_DEBUG	Used for problem determination and may be irrelevant to anyone but the bundle developer.
LOG_ERROR	Indicates the bundle or service may not be functional. Action should be taken to correct this situation.
LOG_INFO	May be the result of any change in the bundle or service and does not indicate a problem.
LOG_WARNING	A bundle or service is still functioning but may experience problems in the future because of the warning condition.

Table 4   *Log Levels*

6.4   Log Reader Service

The Log Reader Service maintains a list of LogEntry objects called the *log*. It is a service that bundle developers can use to retrieve information contained in this log, and receive notifications about LogEntry objects while they are created through the Log Service.

The size of the log is implementation-specific, and it determines how far into the past the log entries go. Additionally, not all log entries may be recorded in the log in order to save space. In particular, LOG_DEBUG log entries may not be recorded. Note that this rule is implementation-dependent; some implementations may allow a configurable policy to ignore certain LogEntry object types.

The LogReaderService interface defines these methods for retrieving log entries.

- getLog() – This method retrieves past log entries as an enumeration with the most recent entry first.
- addLogListener(LogListener) – This method is used to subscribe to the Log Reader Service in order to receive log messages as they occur. Unlike the previously recorded log entries, all log messages must be sent to subscribers of the Log Reader Service as they are recorded.

A subscriber to the Log Reader Service must implement the LogListener interface.

After a subscription to Log Reader Service has been started, the subscriber's LogListener.logged method must be called with a LogEntry object for the message, each time a message is logged.

The LogListener interface defines this method.

- logged(LogEntry) – This method is called for each LogEntry object created. A Log Reader Service implementation must not filter entries, as it is allowed to do for its log. A LogListener object should see all LogEntry objects that are created.

6.4.1 Log Entry

The LogEntry interface abstracts a log entry. It is a record of the information that was passed when an event was logged, and consists of a super set of information which can be passed through the LogService methods. The LogEntry interface defines these methods to retrieve information related to LogEntry objects:

- getBundle() – This method returns the bundle related to a Log-Entry object.
- getException() – This method returns the exception related to a LogEntry object.
- getLevel() – This method returns the severity level related to a LogEntry object.
- getMessage() – This method returns the message related to a LogEntry object.
- getServiceReference() –This method returns the ServiceReference of the service related to a LogEntry object.
- getTime() – This method returns the time that the log entry was created according to the system time.

6.5 Mapping of Events

Implementations of a Log Service must log Framework-generated events and map the information to a LogEntry object in a consistent way. Framework events must be treated exactly the same as other logged events and distributed to all LogListener objects that are associated with the Log Reader Service. The following sections define the mapping for the three different event types.

6.5.1 Bundle Events Mapping

A Bundle Event is mapped to a LogEntry object according to Table 5, "Mapping of Bundle Events to Log Entries," on page 154.

Log Entry method	Information about Bundle Event
getLevel()	LOG_INFO
getBundle()	Identifies the bundle to which the event happened: in other words, the bundle that was installed, started, stopped, updated, or uninstalled. This identification is obtained by calling getBundle() on the BundleEvent object.
getException()	null
getServiceReference()	null
getMessage()	The message depends on the event type: • INSTALLED – "BundleEvent INSTALLED" • STARTED – "BundleEvent STARTED" • STOPPED – "BundleEvent STOPPED" • UPDATED – "BundleEvent UPDATED" • UNINSTALLED – "BundleEvent UNINSTALLED"

Table 5　　　*Mapping of Bundle Events to Log Entries*

6.5.2　Service Events Mapping

A Service Event is mapped to a LogEntry object according to Table 6, "Mapping of Service Events to Log Entries," on page 154..

Log Entry method	Information about Service Event
getLevel()	LOG_INFO, except for the ServiceEvent.MODIFIED event. This event happens frequently and contains relatively little information. It is therefore logged with a level of LOG_DEBUG.
getBundle()	Identifies the bundle that registered the service associated with this event. It is obtained by calling getServiceReference().getBundle() on the ServiceEvent object.
getException()	null

Table 6　　　*Mapping of Service Events to Log Entries*

Log Entry method	Information about Service Event
getServiceReference()	Identifies a reference to the service associated with the event. It is obtained by calling getServiceReference() on the ServiceEvent object.
getMessage()	This message depends on the actual event type. The messages are mapped as follows: • REGISTERED – "ServiceEvent REGISTERED" • MODIFIED – "ServiceEvent MODIFIED" • UNREGISTERING – "ServiceEvent UNREGISTERING"

Table 6   *Mapping of Service Events to Log Entries*

6.5.3   Framework Events Mapping

The Framework can generate two distinctly different events. The first is the STARTED event that indicates that the Framework has finished initializing. This must be done according to Table 7, "Mapping of Framework STARTED Event to Log Entries," on page 155.

Log Entry method	Information about Framework STARTED event
getLevel()	LOG_INFO
getBundle()	Identifies the system bundle. It is obtained by calling get-Bundle() on the FrameworkEvent object.
getException()	null
getServiceReference()	null
getMessage()	"FrameworkEvent STARTED"

Table 7   *Mapping of Framework STARTED Event to Log Entries*

The second event is the ERROR event and indicates that some error occurred. This even is mapped according to Table 8, "Mapping of Framework ERROR Event to Log Entries," on page 155.

Log Entry method	Information about Framework ERROR event
getLevel()	LOG_ERROR
getBundle()	Identifies the bundle associated with the error. It is obtained by calling getBundle() on the FrameworkEvent object.

Table 8   *Mapping of Framework ERROR Event to Log Entries*

Log Entry method	Information about Framework ERROR event
getException()	Identifies the exception associated with the error. It is obtained by calling getThrowable() on the FrameworkEvent object.
getServiceReference()	null
getMessage()	"FrameworkEvent ERROR"

Table 8   *Mapping of Framework ERROR Event to Log Entries*

6.6   Security

The Log Service should only be implemented by trusted bundles. This bundle requires ServicePermission[REGISTER] for the LogService and LogReaderService interfaces. Virtually all bundles should get ServicePermission[GET] for these interfaces.

6.7   Changes Since Release 1.0

- The change to the Log Service 1.1 is to log ServiceEvent.MOD-IFIED at LogService.LOG_DEBUG level instead of LOG_INFO level.
- The message text of the different Framework events is defined.

6.8   org.osgi.service.log

The OSGi Log Service Package. Specification Version 1.1.

Bundles wishing to use this package must list the package in the Import-Package header of the bundle's manifest. For example:

```
Import-Package: org.osgi.service.log; specification-
version=1.1
```

Class Summary

Interfaces

LogEntry	Provides methods to access the information contained in an individual Log Service log entry.
LogListener	Subscribes to LogEntry objects from the LogReaderService.

Class Summary

LogReaderService	Provides methods to retrieve LogEntry objects from the log.
LogService	Provides methods for bundles to write messages to the log.

6.8.1 LogEntry

public interface LogEntry

Provides methods to access the information contained in an individual Log Service log entry.

A LogEntry object may be acquired from the LogReaderService.getLog method or by registering a LogListener object.

See Also: getLog(), LogListener

6.8.1.1 Methods

public Bundle **getBundle**()

Returns the bundle that created this LogEntry object.

Returns: The bundle that created this LogEntry object; null if no bundle is associated with this LogEntry object.

public java.lang.Throwable **getException**()

Returns the exception object associated with this LogEntry object.

Returns: Throwable object of the exception associated with this LogEntry; null if no exception is associated with this LogEntry objcet.

public int **getLevel**()

Returns the severity level of this LogEntry object.

This is one of the severity levels defined by the LogService interface.

Returns: Severity level of this LogEntry object.

See Also: LOG_ERROR, LOG_WARNING, LOG_INFO, LOG_DEBUG

public java.lang.String **getMessage**()

Returns the human readable message associated with this LogEntry object.

Returns: String containing the message associated with this LogEntry object.

public ServiceReference **getServiceReference**()

Returns the ServiceReference object for the service associated with this LogEntry object.

Returns:　　ServiceReference object for the service associated with this LogEntry object; null if no ServiceReference object was provided.

public long **getTime**()

Returns the value of currentTimeMillis() at the time this LogEntry object was created.

Returns:　　The system time in milliseconds when this LogEntry object was created.

See Also:　　System.currentTimeMillis()

6.8.2　LogListener

public interface LogListener extends java.util.EventListener

All Superinterfaces:　　java.util.EventListener

Subscribes to LogEntry objects from the LogReaderService.

A LogListener object may be registered with the Log Reader Service using the LogReaderService.addLogListener method. After the listener is registered, the logged method will be called for each LogEntry object created. The LogListener object may be unregistered by calling the LogReaderService.removeLogListener method.

See Also:　　LogReaderService, LogEntry, addLogListener(LogListener), removeLogListener(LogListener)

6.8.2.1　Methods

public void **logged**(LogEntry entry)

Listener method called for each LogEntry object created.

As with all event listeners, this method should return to its caller as soon as possible.

Parameters:　　entry - A LogEntry object containing log information.

See Also:　　LogEntry

6.8.3　LogReaderService

public interface LogReaderService

Provides methods to retrieve LogEntry objects from the log.

There are two ways to retrieve LogEntry objects:

- The primary way to retrieve LogEntry objects is to register a Log-Listener object whose LogListener.logged method will be called for each entry added to the log.
- To retrieve past LogEntry objects, the getLog method can be called which will return an Enumeration of all LogEntry objects in the log.

See Also: LogEntry, LogListener, logged(LogEntry)

6.8.3.1   **Methods**

public void **addLogListener**(LogListener listener)

Subscribes to LogEntry objects.

This method registers a LogListener object with the Log Reader Service. The LogListener.logged(LogEntry) method will be called for each LogEntry object placed into the log.

When a bundle which registers a LogListener object is stopped or otherwise releases the Log Reader Service, the Log Reader Service must remove all of the bundle's listeners.

Parameters: listener - A LogListener object to register; the LogListener object is used to receive LogEntry objects.

See Also: LogListener, LogEntry, logged(LogEntry)

public java.util.Enumeration **getLog**()

Returns an Enumeration of all LogEntry objects in the log.

Each element of the enumeration is a LogEntry object, ordered with the most recent entry first. Whether the enumeration is of all Log-Entry objects since the Log Service was started or some recent past is implementation-specific. Also implementation-specific is whether informational and debug LogEntry objects are included in the enumeration.

public void **removeLogListener**(LogListener listener)

Unsubscribes to LogEntry objects.

This method unregisters a LogListener object from the Log Reader Service.

Parameters: listener - A LogListener object to unregister.

See Also: LogListener

6.8.4 LogService

public interface LogService

Provides methods for bundles to write messages to the log.

LogService methods are provided to log messages; optionally with a ServiceReference object or an exception.

Bundles must log messages in the OSGi environment with a severity level according to the following hierarchy:

1. LOG_ERROR

2. LOG_WARNING

3. LOG_INFO

4. LOG_DEBUG

6.8.4.1 Fields

public static final int **LOG_DEBUG**

A debugging message (Value 4).

This log entry is used for problem determination and may be irrelevant to anyone but the bundle developer.

public static final int **LOG_ERROR**

An error message (Value 1).

This log entry indicates the bundle or service may not be functional.

public static final int **LOG_INFO**

An informational message (Value 3).

This log entry may be the result of any change in the bundle or service and does not indicate a problem.

public static final int **LOG_WARNING**

A warning message (Value 2).

This log entry indicates a bundle or service is still functioning but may experience problems in the future because of the warning condition.

6.8.4.2 Methods

public void **log**(int level, java.lang.String message)

Logs a message.

The ServiceReference field and the Throwable field of the LogEntry object will be set to null.

Parameters: level - The severity of the message; one of the four defined severities.

message - Human readable string describing the condition.

See Also: LOG_ERROR, LOG_WARNING, LOG_INFO, LOG_DEBUG

public void **log**(int level, java.lang.String message,
    java.lang.Throwable exception)

Logs a message with an exception.

The ServiceReference field of the LogEntry object will be set to null.

Parameters: level - The severity of the message; one of the four defined severities.

message - The human readable string describing the condition.

exception - The exception that reflects the condition.

See Also: LOG_ERROR, LOG_WARNING, LOG_INFO, LOG_DEBUG

public void **log**(ServiceReference sr, int level, java.lang.String
    message)

Logs a message associated with a specific ServiceReference object.

The Throwable field of the LogEntry will be set to null.

Parameters: sr - The ServiceReference object of the service that this message is associated with.

level - The severity of the message; one of the four defined severities.

message - Human readable string describing the condition.

See Also: LOG_ERROR, LOG_WARNING, LOG_INFO, LOG_DEBUG

public void **log**(ServiceReference sr, int level, java.lang.String
    message, java.lang.Throwable exception)

Logs a message with an exception associated and a Service-Reference object.

Parameters: sr - The ServiceReference object of the service that this message is associated with.

level - The severity of the message; one of the four defined severities.

message - Human readable string describing the condition.

exception - The exception that reflects the condition.

See Also: LOG_ERROR, LOG_WARNING, LOG_INFO, LOG_DEBUG

7 Http Service Specification

Version 1.1

7.1 Introduction

An OSGi environment normally provides users with access to services on the Internet and other networks. This access allows users to remotely retrieve information from, and send control to, services in an OSGi environment using a standard web browser.

Bundle developers typically need to develop communication and user-interface solutions for standard technologies such as HTTP, HTML, XML, and servlets.

The Http Service supports two standard techniques for this purpose:

- *Servlets* – A servlet is a Java object which implements the Java Servlet API. Registering a servlet in the Framework gives it control over some part of the Http Service URI namespace.
- *Resources* – Registering a resource allows HTML files, image files, and other static resources to be made visible in the Http Service URI namespace by the requesting bundle.

Implementations of the Http Service can be based on:

- [13] *HTTP 1.0 Specification RFC-1945*
- [14] *HTTP 1.1 Specification RFC-2616*

Alternatively, implementations of this service can support other protocols if these protocols can conform to the semantics of the javax.servlet API. This additional support is necessary because the Http Service is closely related to [15] *Java Servlet Technology*. Http Service implementations must support at least version 2.1 of the Java Servlet API.

7.1.1 Entities

This specification defines the following interfaces which a bundle developer can implement collectively as an Http Service or use individually:

- HttpContext – Allows bundles to provide information for a servlet or resource registration.

- HttpService – Allows other bundles in the Framework to dynamically register and unregister resources and servlets into the Http Service URI namespace.
- NamespaceException – Is thrown to indicate an error with the caller's request to register a servlet or resources into the Http Service URI namespace.

Figure 15   *Http Service Overview Diagram*

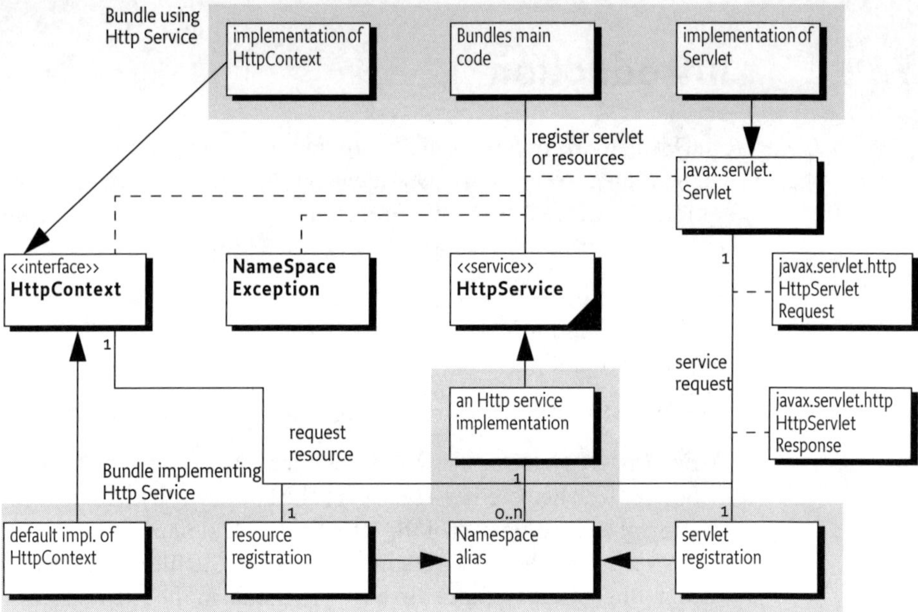

7.2   Registering Servlets

javax.servlet.Servlet objects can be registered with the Http Service by using the HttpService interface. For this purpose, the HttpService interface defines the method registerServlet(String, Servlet, Dictionary, HttpContext).

For example, if the Http Service implementation is listening to port 80 on the machine www.acme.com and the Servlet object is registered with the name "/servlet", then the Servlet object's service method is called when the following URL is used from a web browser:

```
http://www.acme.com/servlet?name=bugs
```

All Servlet objects and resource registrations share the same namespace. If an attempt is made to register a resource or Servlet object under the same name as a currently registered resource or Servlet object, a NameSpaceException is thrown. See *Mapping HTTP Requests to Servlet and Resource Registrations* on page 169 for more information about the handling of the Http Service namespace.

Each Servlet registration must be accompanied with an HttpContext object. This object provides the handling of resources, media typing, and a method to handle authentication of remote requests. See *Authentication* on page 173.

For convenience, a default HttpContext object is provided by the Http Service and can be obtained with createDefaultHttpContext(). Passing a null parameter to the registration method achieves the same effect.

Servlet objects require a ServletContext object. This object provides a number of functions to access the Http Service Java Servlet environment. It is created by the implementation of the Http Service for each unique HttpContext object with which a Servlet object is registered. Thus, Servlet objects registered with the same HttpContext object must also share the same ServletContext object.

Servlet objects are initialized by the Http Service when they are registered and bound to that specific Http Service. The initialization is done by calling the Servlet object's Servlet.init(ServletConfig) method. The ServletConfig parameter provides access to the initialization parameters specified when the Servlet object was registered.

Therefore, the same Servlet instance must not be reused for registration with another Http Service, nor can it be registered under multiple names. Unique instances are required for each registration.

The following example code demonstrates the use of the registerServlet method:

```
Hashtable initparams = new Hashtable();
initparams.put( "name", "value" );

Servlet myServlet = new HttpServlet() {
  String    name = "<not set>";

  public void init( ServletConfig config ) {
    this.name = (String)
      config.getInitParameter( "name" );
  }
```

```
      public void doGet(
        HttpServletRequest req,
        HttpServletResponse rsp
      ) throws IOException {
        rsp.setContentType( "text/plain" );

        req.getWriter().println( this.name );
      }
    };

    getHttpService().registerServlet(
      "/servletAlias",
      myServlet,
      initparams,
      null // use default context
    );
    // myServlet has been registered
    // and its init method has been called. Remote
    // requests are now handled and forwarded to
    // the servlet.
    ...
    getHttpService().unregister("/servletAlias");
    // myServlet has been unregistered and its
    // destroy method has been called
```

The example registers the servlet, myServlet, at alias: /servletAlias.
Future requests for http://www.acme.com:80/servletAlias, maps to
the servlet, myServlet, whose service method is called to process the
request. (The service method is called in the HttpServlet base class
and dispatched to a doGet, doPut, doPost, doOptions, doTrace or
doDelete call depending on the HTTP request method used.)

7.3 Registering Resources

A resource is a file containing images, static HTML pages, sounds,
movies, applets, and the like. Resources do not require any handling
from the bundle. They are transferred directly from their source,
usually the JAR file that contains the code for the bundle, to the
requestor using HTTP.

Resources could be handled by Servlet objects as explained in *Registering Servlets* on page 164. Transferring a resource over HTTP, however, would require very similar Servlet objects for each bundle. To
prevent this redundancy, resources can be registered directly with

the Http Service via the HttpService interface. This HttpService interface defines the registerResources(String, String, HttpContext) method for registering a resource into the Http Service URI namespace.

The first parameter is the external alias under which the resource is registered with the Http Service. The second parameter is an internal prefix to map this resource to the bundle's namespace. When a request comes in, the HttpService object must remove the external alias from the URI and replace it with the internal prefix, and then call the getResource(String) method with this new name on the associated HttpContext object. The HttpContext object is further used to get the MIME type of the resource and to authenticate the request.

Resources are returned as a java.net.URL object. The Http Service must read from this URL object and transfer the content to the initiator of the HTTP request.

This return type was chosen because it matches the return type of the java.lang.Class.getResource(String resource) method. This method can retrieve resources directly from the same place as the one from which the class was loaded – often a package directory in the JAR file of the bundle. This method makes it very convenient to retrieve resources from the bundle that are contained in the package.

The following example code demonstrates the use of the register Resources method:

```java
package com.acme;
...
HttpContext context = new HttpContext() {
  public boolean handleSecurity(
    HttpServletRequest request,
    HttpServletResponse response
  ) throws IOException {
    return true;
  }

  public URL getResource(String name) {
    return getClass().getResource(name);
  }

  public String getMimeType(String name) {
    return null;
  }
};
```

```
httpService.registerResources (
   "/files",
   "www",
   context
);
...
getHttpService().unregister("/files");
```

This example registers the alias, /files, on the Http Service. Requests for resources below this name space are transferred to the HttpContext object with an internal name of www/<name>. This example uses the Class.get
Resource(String) method. Because the internal name does not start with a
"/", it must map to a resource in the "com/acme/www" directory of the JAR file. If the internal name did start with a "/", the package name would not have to be prefixed and the JAR file would be searched from the root. Consult the java.lang.Class.getResource(String) method for more information.

In the example, a request for http://www.acme.com/files/myfile.html must thus map to the name "/com/acme/www/myfile.html" which is in the bundle's JAR file.

More sophisticated implementations of the getResource(String) method could filter the input name, restricting the resources that may be returned or map the input name onto the file system (if the security implications of this choice would be acceptable).

Alternatively, the resource registration could have used a default HttpContext object, as demonstrated in the following call to registerResources:

```
getHttpService().registerResources(
   "/files",
   "/com/acme/www",
   null
);
```

In this case, the Http Service implementation would call the createDefaultHttpContext() method and use its return value as the HttpContext argument for the registerResources method. The default implementation must map the resource request to the bundle's resource, using

Bundle.getResource(String). In the case of the previous example, however, the internal name must now specify the full path to the directory containing the resource files in the JAR file; no automatic prefixing of the package name is done.

The getMime(String) implementation of the default HttpContext object should return a reasonable mapping. Its handleSecurity(HttpServlet Request,HttpServletResponse) may implement an authentication mechanism that is implementation-dependent.

7.4 Mapping HTTP Requests to Servlet and Resource Registrations

When an HTTP request comes in from a client, the Http Service checks to see if the requested URI matches any registered aliases. A URI matches only if the path part of the URI is exactly the same string; matching is case sensitive.

If it does match, then there is a matching registration.

- If the registration corresponds to a servlet, the authorization is verified by calling the handleSecurity method of the associated HttpContext object (see *Authentication* on page 173). If the request is authorized, the servlet must be called by its service method to complete the HTTP request.
- If the registration corresponds to a resource, the authorization is verified by calling the handleSecurity method of the associated HttpContext object (see *Authentication* on page 173). If the request is authorized, a target resource name is constructed by substituting the alias name from the registration with the internal name from the registration.
  The target resource name must be passed to the getResource method of the associated HttpContext object.
  If the returned URL object is not null, the Http Service must return the contents of the URL to the client completing the HTTP request. The translated target name (as opposed to the original requested URI) must also be used as the argument to HttpContext.getMimeType.
  If the returned URL object is null, the Http Service continues as if there was no match.
- If there is no match, the Http Service must attempt to match substrings of the requested URI to registered aliases. The substrings of the requested URI are selected by removing the last "/" and everything to its right.

The Http Service must repeat this process until either a match is found or the substring is an empty string. If the substring is empty and the alias "/" is registered, the request is considered to match the alias "/". Otherwise the Http Service must return HttpServletResponse.SC_NOT_FOUND(404) to the client.

For example, an HTTP request comes in with a request URI of "/fudd/bugs/foo.txt", and the only registered alias is "/fudd", a search for "/fudd/bugs/foo.txt", will not match an alias; therefore the Http Service with search for the alias "/fudd/bugs", and the alias "/fudd". The latter search will result in a match and the matched alias registration must be used.

Registrations for identical aliases are not allowed; if a bundle registers the alias "/fudd", and another bundle tries to register the exactly the same alias, the second caller must receive a NamespaceException and its resource or servlet must *not* be registered. It could, however, register a similar alias – for example, "/fudd/bugs" – as long as no other registration for this alias already exists.

The following table shows some examples of the usage of the name space.

Alias	Internal Name	URI	getResource Parameter
/	(empty string)	/fudd/bugs	fudd/bugs
/	/fudd	/bugs	/fudd/bugs
/fudd	(empty string)	/fudd/bugs	/bugs
/fudd	fudd	/fudd/bugs	fudd/bugs
/fudd	/tmp	/fudd/bugs	/tmp/bugs
/fudd	/tmp	/fudd/bugs/x.gif	/tmp/bugs/x.gif
/fudd/bugs	tmp	/fudd/bugs/x.gif	tmp/x.gif

Table 9   *Examples of Namespace Mapping*

7.5   The Default Http Context Object

The HttpContext object in the first example demonstrates simple implementations of the HttpContext interface methods. Alternatively, the example could have used a default HttpContext object, as demonstrated in the following call to registerServlet:

```
getHttpService().registerServlet(
    "/servletAlias",
    myServlet,
```

```
      initparams,
      null
  );
```

In this case, the Http Service implementation must call createDefault
HttpContext and use the return value as the HttpContext argument.

If the default HttpContext object, and thus the ServletContext
object, is to be shared by multiple servlet registrations, the previous
servlet registration example code needs to be changed to use the
same default HttpContext object. This change is demonstrated in
the next example:

```
HttpContext defaultContext =
   getHttpService().createDefaultHttpContext();

getHttpService().registerServlet(
   "/servletAlias",
   myServlet,
   initparams,
   defaultContext
);

// defaultContext can be reused
// for further servlet registrations
```

7.6   MIME Types

MIME (Multipurpose Internet Mail Extension) defines an extensive
set of headers and procedures to encode binary messages in US-
ASCII mails. For an overview of all the related RFCs, consult [16]
MIME Multipurpose Internet Mail Extension.

An import aspect of this extension is the type (file format) mechanism of the binary messages. The type is defined by a string containing a general category (text, application, image, audio and video,
multipart, and message) followed by a "/" and a specific media type:
for example, "text/html" for HTML formatted text files. A MIME
type string can be followed by further specifiers by separating a
key=value pair with a ';'. These specifiers can be used, for example, to
define character sets:

```
text/plan ; charset=iso-8859-1
```

The Internet Assigned Number Authority (IANA) maintains a set of defined MIME media types. This list can be found at [17] *Assigned MIME Media Types*. MIME media types are extendable, and when any part of the type starts with the prefix "x-", it is assumed to be vendor-specific and can be used for testing. New types can be registered at [18] *Registration Procedures for new MIME media types*.

HTTP bases its media typing on the MIME RFCs. The "Content-Type" header should contain a MIME media type so that the browser can recognize the type and format the content correctly.

The source of the data must define the MIME media type for each transfer. Most operating systems do not support types for files, but use conventions based on file names: for example, the last part of the file name after the last ".". This extension is then mapped to a media type via a table.

Implementations of the Http Service should have a reasonable default of mapping common extension to media types based on file extensions.

Extension	MIME media type	Description
.jpg .jpeg	image/jpeg	JPEG Files
.gif	image/gif	GIF Files
.css	text/css	Cascading Style Sheet Files
.txt	text/plain	Text Files
.wml	text/vnd.wap.wml	Wireless Access Protocol (WAP) Mark Language
.htm .html	text/html	Hyper Text Markup Language
.wbmp	image/ vnd.wap.wbmp	Bitmaps for WAP

Table 10   *Sample Extension to MIME Media Mapping*

Only the bundle developer, however, knows exactly which files have what media type. The HttpContext interface can therefore be used to map this knowledge to the media type. The HttpContext class has the following method for this: getMimeType(String).

The implementation of this method should inspect the file name and use its internal knowledge to map this name to a MIME media type.

Simple implementation can extract the extension and look up this extension in a table.

Returning null from this method allows the Http Service implementation to use its default mapping mechanism.

7.7 Authentication

The Http Service has separated the authentication and authorization of a request from the execution of the request. This separation allows bundles to use available Servlet sub-classes while still providing bundle specific authentication and authorization of the requests.

Prior to servicing each incoming request, the Http Service calls the handleSecurity(HttpServletRequest, HttpServletResponse) method on the HttpContext object that is associated with the request URI. This method controls whether the request is processed in the normal manner or an authentication error is returned.

If an implementation wants to authenticate the request, it can use the authentication mechanisms of HTTP. See [19] *RFC 2617: HTTP Authentication: Basic and Digest Access Authentication.* These mechanisms normally interpret the headers and decide if the user identity is available, and if it is, whether that user has authenticated itself correctly.

There are many different ways of authenticating users, and the handleSecurity method on the HttpContext object can use whatever method it requires. If the method returns true, the request must continue to be processed using the potentially modified HttpServletRequest and HttpServletResponse objects. If the method returns false, the request must *not* be processed.

A common standard for HTTP is the basic authentication scheme that is not secure when used with HTTP. Basic authentication passes the password in base 64 encoded strings that are trivial to decode into clear text. Secure transport protocols like HTTPS use SSL to hide this information. With these protocols basic authentication is secure.

Using basic authentication requires the following steps:

- If no Authorization header is set in the request, the method should set the WWW-Authenticate header in the response. This header indicates the desired authentication mechanism and the realm. As a simple example: WWW-Authenticate: Basic realm="ACME"
The header should be set with the response object that is given as a parameter to the handleSecurity method. The handleSecurity

method should set the status to HttpServletRe-
sponse.SC_UNAUTHORIZED (401) and return false.

- Secure connections can be verified with the ServletRequest.
getScheme() method. This method returns, for example, "https"
for an SSL connection. Then the handleSecurity method can use
this and other information to decide if the connection's security
level is acceptable. If not, the handleSecurity method should set
the status to HttpServletResponse.SC_FORBIDDEN (403) and
return false.

- Next, the request must be authenticated. When basic authentication is used, the Authorization header is available in the
request and should be parsed to find the user and password. See
[19] *RFC 2617: HTTP Authentication: Basic and Digest Access Authentication* for more information.
If the user cannot be authenticated, the status of the response
object should be set to HttpServletRe-
sponse.SC_UNAUTHORIZED (401) and return false.

- The authentication mechanism that is actually used and the
identity of the authenticated user can be of interest to the Servlet
object. Therefore, the implement of the handleSecurity method
should then set this information in the request object using the
ServletRequest.setAttribute method. This specification has
defined a number of OSGi-specific attribute names for this
purpose:
  - AUTHENTICATION_TYPE - Specifies the scheme used in
  authentication. A Servlet may retrieve the value of this
  attribute by calling the HttpServletRequest.getAuthType
  method. This attribute name is org.osgi.service.http.authen-
  tication.type.
  - REMOTE_USER - Specifies the name of the authenticated user.
  A Servlet may retrieve the value of this attribute by calling the
  HttpServletRequest.getRemoteUser method. This attribute
  name is org.osgi.service.http.authentication.remote.user.
  - AUTHORIZATION - If a User Admin service is available in the
  environment, then the handleSecurity method should set this
  attribute with the Authorization object obtained from the
  User Admin service. Such an object encapsulates the authentication of its remote user. A Servlet may retrieve the value of
  this attribute by calling ServletRequest.getAttribute(Http-
  Context.AUTHORIZATION). This header name is org.osgi.ser-
  vice.useradmin.authorization.

- Once the request is authenticated and any attributes are set, the
handleSecurity method should return true. This return indicates
to the Http Service that the request is authorized and processing
may continue. If the request is for a Servlet, the Http Service must
then call the service method on the Servlet object.

7.8   Security

This section only applies when executing in an OSGi environment which is enforcing Java permissions.

7.8.1   Accessing Resources in Bundles

The Http Service must be granted AdminPermission so that bundles may use a default HttpContext object. This is necessary because the implementation of the default HttpContext object must call Bundle.getResource to access the resources of a bundle and this method requires the caller to have AdminPermission.

Any bundle may access resources in its own bundle by calling Class.getResource. This is a privileged operation. The resulting URL object may then be passed to the Http Service as the result of a HttpContext.getResource call. No further permission checks are performed when accessing bundle resource URL objects so the Http Service does not need to be granted any additional permissions.

7.8.2   Accessing Other Types of Resources

In order to open resources that were not registered using the default HttpContext object, the Http Service must be granted sufficient privileges to access these resources. For example, if the getResource method of the registered HttpContext object returns a file URL, the Http Service requires the corresponding FilePermission to read the file. Likewise, if the getResource method of the registered HttpContext object returns an http URL, the Http Service requires the corresponding SocketPermission to connect to the resource.

This means that, in most cases, Http Service should be a privileged service, granted sufficient permission to serve any bundle's resources, no matter where these resources are located. Therefore, the Http Service must capture the AccessControlContext object of the bundle registering resources or a servlet and then use the captured AccessControlContext object when accessing resources returned by the registered HttpContext object. This prevents a bundle from registering resources that it does not have permission to access.

Therefore, the Http Service should follow a scheme like the next example. When a resource or servlet is registered, it should capture the context.

```
AccessControlContext acc =
    AccessController.getContext();
```

When a URL returned by the getResource method of the associated HttpContext object is called, the Http Service must do so in a doPrivileged construct using the AccessControlContext object of the registering bundle:

```
doPrivileged(new PrivilegedExceptionAction() {
  public Object run() throws Exception {
    ...
  }
}, acc);
```

The Http Service must only use the captured AccessControlContext when accessing resource URL objects. Servlet and HttpContext objects must use a doPrivileged construct in their implementations when performing privileged operations.

7.9 Configuration Properties

If the Http Service does not have its port values configured through some other means, the Http Service implementation should use the following properties to determine the port values upon which to listen.

The following OSGi environment properties are used to specify default HTTP ports:

- org.osgi.service.http.port – This property specifies the port used for servlets and resources accessible via HTTP. The default value for this property is 80.
- org.osgi.service.http.port.secure – This property specifies the port used for servlets and resources accessible via HTTPS. The default value for this property is 443.

7.10 org.osgi.service.http

The OSGi Http Service Package. Specification Version 1.1.

Bundles wishing to use this package must list the package in the Import-Package header of the bundle's manifest. For example:

```
Import-Package: org.osgi.service.http; specification-
version=1.1
```

Class Summary

Interfaces

HttpContext	This interface defines methods that the Http Service may call to get information about a registration.
HttpService	The Http Service allows other bundles in the OSGi environment to dynamically register resources and servlets into the URI namespace of Http Service.

Exceptions

NamespaceException	A NamespaceException is thrown to indicate an error with the caller's request to register a servlet or resources into the URI namespace of the Http Service.

7.10.1 HttpContext

public interface HttpContext

This interface defines methods that the Http Service may call to get information about a registration.

Servlets and resources may be registered with an Http Context object; if no Http Context object is specified, a default Http Context object is used. Servlets that are registered using the same Http-Context object will share the same Servlet Context object.

This interface is implemented by users of the Http Service.

7.10.1.1 Fields

public static final java.lang.String **AUTHENTICATION_TYPE**

Http Servlet Request attribute specifying the scheme used in authentication. The value of the attribute can be retrieved by Http-Servlet Request.get Auth Type. This attribute name is org.osgi.service.http.authentication.type.

Since: 1.1

public static final java.lang.String **AUTHORIZATION**

Http Servlet Request attribute specifying the Authorization object obtained from the org.osgi.service.useradmin.User Admin service. The value of the attribute can be retrieved by Http Servlet-Request.get Attribute(Http Context.AUTHORIZATION). This attribute name is org.osgi.service.useradmin.authorization.

Since: 1.1

public static final java.lang.String **REMOTE_USER**

Http Servlet Request attribute specifying the name of the authenticated user. The value of the attribute can be retrieved by HttpServletRequest.getRemoteUser. This attribute name is org.osgi.service.http.authentication.remote.user.

Since: 1.1

7.10.1.2   **Methods**

public java.lang.String **getMimeType**(java.lang.String name)

Maps a name to a MIME type. Called by the Http Service to determine the MIME type for the name. For servlet registrations, the Http Service will call this method to support the Servlet Context method getMimeType. For resource registrations, the Http Service will call this method to determine the MIME type for the Content-Type header in the response.

Parameters: name - determine the MIME type for this name.

Returns: MIME type (e.g. text/html) of the name or null to indicate that the Http Service should determine the MIME type itself.

public java.net.URL **getResource**(java.lang.String name)

Maps a resource name to a URL.

Called by the Http Service to map a resource name to a URL. For servlet registrations, Http Service will call this method to support the Servlet Context methods getResource and getResourceAsStream. For resource registrations, Http Service will call this method to locate the named resource. The context can control from where resources come. For example, the resource can be mapped to a file in the bundle's persistent storage area via bundleContext.getDataFile(name).toURL() or to a resource in the context's bundle via getClass().getResource(name)

Parameters: name - the name of the requested resource

Returns: URL that Http Service can use to read the resource or null if the resource does not exist.

public boolean **handleSecurity**(
    javax.servlet.http.HttpServletRequest request,
    javax.servlet.http.HttpServletResponse response)
    throws java.io.IOException

Handles security for the specified request.

The Http Service calls this method prior to servicing the specified request. This method controls whether the request is processed in the normal manner or an error is returned.

If the request requires authentication and the Authorization header in the request is missing or not acceptable, then this method should set the WWW-Authenticate header in the response object, set the status in the response object to Unauthorized(401) and return false. See also RFC 2617: *HTTP Authentication: Basic and Digest Access Authentication* (available at http://www.ietf.org/rfc/rfc2617.txt).

If the request requires a secure connection and the get Scheme method in the request does not return 'https' or some other acceptable secure protocol, then this method should set the status in the response object to Forbidden(403) and return false.

When this method returns false, the Http Service will send the response back to the client, thereby completing the request. When this method returns true, the Http Service will proceed with servicing the request.

If the specified request has been authenticated, this method must set the AUTHENTICATION_TYPE request attribute to the type of authentication used, and the REMOTE_USER request attribute to the remote user (request attributes are set using the set Attribute method on the request). If this method does not perform any authentication, it must not set these attributes.

If the authenticated user is also authorized to access certain resources, this method must set the AUTHORIZATION request attribute to the Authorization object obtained from the org.osgi.service.useradmin.User Admin service.

The servlet responsible for servicing the specified request determines the authentication type and remote user by calling the get-Auth Type and get Remote User methods, respectively, on the request.

Parameters: request - the HTTP request

                    response - the HTTP response

Returns: true if the request should be serviced, false if the request should not be serviced and Http Service will send the response back to the client.

Throws: java.io.IOException - may be thrown by this method. If this occurs, the Http Service will terminate the request and close the socket.

7.10.2   **HttpService**

public interface HttpService

The Http Service allows other bundles in the OSGi environment to dynamically register resources and servlets into the URI namespace of Http Service. A bundle may later unregister its resources or servlets.

See Also: HttpContext

7.10.2.1   **Methods**

public HttpContext **createDefaultHttpContext**()

Creates a default Http Context for registering servlets or resources with the HttpService, a new Http Context object is created each time this method is called.

The behavior of the methods on the default Http Context is defined as follows:

getMimeType

> Does not define any customized MIME types for the Content-Type header in the response, and always returns null.

handleSecurity

> Performs implementation-defined authentication on the request.

getResource

> Assumes the named resource is in the context bundle; this method calls the context bundle's Bundle.getResource method, and returns the appropriate URL to access the resource. On a Java runtime environment that supports permissions, the Http Service needs to be granted the org.osgi.framework.AdminPermission.

Returns: a default Http Context object.

Since: 1.1

public void **registerResources**(java.lang.String alias,
     java.lang.String name, HttpContext context)
     throws NamespaceException,
     java.lang.IllegalArgumentException

Registers resources into the URI namespace.

The alias is the name in the URI namespace of the Http Service at which the registration will be mapped. An alias must begin with slash ('/') and must not end with slash ('/'), with the exception that an alias of the form "/" is used to denote the root alias. The name parameter must also not end with slash ('/'). See the specification text for details on how HTTP requests are mapped to servlet and resource registrations.

For example, suppose the resource name /tmp is registered to the alias /files. A request for /files/foo.txt will map to the resource name /tmp/foo.txt.

```
httpservice.registerResources("/files",
  "/tmp",
  context);
```

The Http Service will call the Http Context argument to map resource names to URLs and MIME types and to handle security for requests. If the Http Context argument is null, a default Http-Context is used (see createDefaultHttpContext()).

Parameters: alias - name in the URI namespace at which the resources are registered

name - the base name of the resources that will be registered

context - the Http Context object for the registered resources, or null if a default Http Context is to be created and used.

Throws: NamespaceException - if the registration fails because the alias is already in use.

java.lang.IllegalArgumentException - if any of the parameters are invalid

```
public void registerServlet(java.lang.String alias,
    javax.servlet.Servlet servlet, java.util.Dictionary initparams,
    HttpContext context)
    throws NamespaceException, javax.servlet.ServletException,
    java.lang.IllegalArgumentException
```

Registers a servlet into the URI namespace.

The alias is the name in the URI namespace of the Http Service at which the registration will be mapped.

An alias must begin with slash ('/') and must not end with slash ('/'), with the exception that an alias of the form "/" is used to denote the root alias. See the specification text for details on how HTTP requests are mapped to servlet and resource registrations.

The Http Service will call the servlet's init method before returning.

```
httpService.registerServlet("/myservlet",
    servlet,
    initparams,
    context);
```

Servlets registered with the same Http Context object will share the same Servlet Context. The Http Service will call the context argument to support the Servlet Context methods get Resource, get-Resource As Stream and get Mime Type, and to handle security for requests. If the context argument is null, a default Http Context object is used (see createDefaultHttpContext()).

Parameters: alias - name in the URI namespace at which the servlet is registered

servlet - the servlet object to register

initparams - initialization arguments for the servlet or null if there are none. This argument is used by the servlet's Servlet Config object.

context - the Http Context object for the registered servlet, or null if a default Http Context is to be created and used.

Throws: NamespaceException - if the registration fails because the alias is already in use.

javax.servlet.ServletException - if the servlet's init method throws an exception, or the given servlet object has already been registered at a different alias.

java.lang.IllegalArgumentException - if any of the arguments are invalid

public void **unregister**(java.lang.String alias)
    throws java.lang.IllegalArgumentException

Unregisters a previous registration done by registerServlet or registerResources.

After this call, the registered alias in the URI namespace will no longer be available. If the registration was for a servlet, HttpService will call the destroy method of the servlet before returning.

If the bundle which performed the registration is stopped or otherwise "unget"s the Http Service without calling unregister(String) then HttpService must automatically unregister the registration. However, if the registration was for a servlet, the destroy method of the servlet will not be called in this case since the bundle may be stopped. unregister(String) must be explicitly called to cause the destroy method of the servlet to be called. This can be done in the BundleActivator.stop(BundleContext) method of the bundle registering the servlet.

Parameters: alias - name in the URI namespace of the registration to unregister

Throws: java.lang.IllegalArgumentException - if there is no registration for the alias or the calling bundle was not the bundle which registered the alias.

7.10.3   NamespaceException

public class NamespaceException extends java.lang.Exception

All Implemented Interfaces: java.io.Serializable

A NamespaceException is thrown to indicate an error with the caller's request to register a servlet or resources into the URI namespace of the Http Service. This exception indicates that the requested alias already is in use.

7.10.3.1   Constructors
public **NamespaceException**(java.lang.String message)

Construct a NamespaceException object with a detail message.

Parameters: message - the detail message

public **NamespaceException**(java.lang.String message,
    java.lang.Throwable exception)

Construct a NamespaceException object with a detail message and a nested exception.

Parameters: message - the detail message

exception - the nested exception

7.10.3.2   Methods
public java.lang.Throwable **getException**()

Returns the nested exception.

Returns: the nested exception or null if there is no nested exception.

7.11   References

[13]   *HTTP 1.0 Specification RFC-1945*
Available at htpp://www.ietf.org/rfc/rfc1945.txt, May 1996

[14]   *HTTP 1.1 Specification RFC-2616*
Available at http://www.ietf.org/rfc/rfc2616.txt, June 1999

[15] *Java Servlet Technology*
Available at http://java.sun.com/products/servlet/index.html

[16] *MIME Multipurpose Internet Mail Extension*
Available at http://www.nacs.uci.edu/indiv/ehood/MIME/
MIME.html

[17] *Assigned MIME Media Types*
ftp://ftp.isi.edu/in-notes/iana/assignments/media-types/media-types

[18] *Registration Procedures for new MIME media types*
Available at http://www.nacs.uci.edu/indiv/ehood/MIME/2048/
rfc2048.html

[19] *RFC 2617: HTTP Authentication: Basic and Digest Access Authentication*
Available at http://www.ietf.org/rfc/rfc2617.txt

8   Device Access Specification

Version 1.1

8.1   Introduction

A service platform is a meeting point for services and devices from
many different vendors: a meeting point where users add and cancel
service subscriptions, newly installed services find their corresponding input and output devices, and device drivers connect to their
hardware.

In an OSGi environment, these activities will dynamically take
place while the Framework is running. Technologies such as USB
and IEEE 1394 explicitly support plugging and unplugging devices
at any time, and wireless technologies are even more dynamic.

This flexibility makes it hard to configure all aspects of an OSGi
environment, particularly those relating to devices. When all of the
possible services and device requirements are factored in, each OSGi
environment will be unique. Therefore, automated mechanisms are
needed that can be extended and customized, in order to minimize
the configuration needs of the OSGi environment.

The Device Access specification supports the coordination of automatic detection and attachment of existing devices on an OSGi environment, facilitates hot-plugging and -unplugging of new devices,
and downloads and installs device drivers on demand.

This specification, however, deliberately does not prescribe any particular device or network technology, and mentioned technologies
are used as examples only. Nor does it specify a particular device discovery method. Rather, this specification focuses on the attachment
of devices supplied by different vendors. It emphasizes the development of standardized device interfaces to be defined in device categories, although no such device categories are defined in this
specification.

8.1.1   Essentials

- *Embedded Devices* – OSGi bundles will likely run in embedded
  devices. This environment implies limited possibility for user

interaction, and low-end devices will probably have resource limitations.

- *Remote Administration* – OSGi environments s must support administration by a remote service provider.
- *Vendor Neutrality* – OSGi-compliant driver bundles will be supplied by different vendors; each driver bundle must be well-defined, documented, and replaceable.
- *Continuous Operation* – OSGi environments will be running for extended periods without being restarted, possibly continuously, requiring stable operation and stable resource consumption.
- *Dynamic Updates* – As much as possible, driver bundles must be individually replaceable without affecting unrelated bundles. In particular, the process of updating a bundle should not require a restart of the whole OSGi environment or disrupt operation of connected devices.

A number of requirements must be satisfied by Device Access implementations in order for it to be OSGi-compliant. Implementations must support the following capabilities:

- *Hot-Plugging* – Plugging and unplugging of devices at any time if the underlying hardware and drivers allow it.
- *Legacy Systems* – Device technologies which do not implement the automatic detection of plugged and unplugged devices.
- *Dynamic Device Driver Loading* – Loading new driver bundles on demand with no prior device-specific knowledge of the Device service.
- *Multiple Device Representations* – Devices to be accessed from multiple levels of abstraction.
- *Deep Trees* – Connections of devices in a tree of mixed network technologies of arbitrary depth.
- *Topology Independence* – Separation of the interfaces of a device from where and how it is attached.
- *Complex Devices* – Multifunction devices and devices that have multiple configurations.

8.1.2 Operation

This specification defines the behavior of a device manager (which is *not* a service as might be expected). This device manager detects registration of Device services and is responsible for associating these devices with an appropriate Driver service. This done with the help of Driver Locator services and the Driver Selector service that allow a device manager to find a Driver bundle and install it.

8.1.3 Entities

The main entities of the Device Access specification are:

- *Device Manager* – The bundle that controls the initiation of the attachment process behind the scenes.
- *Device Category* – Defines how a Driver service and a Device service can cooperate.
- *Driver* – Competes for attaching Device services of its recognized device category. See *Driver Services* on page 193.
- *Device* – A representation of a physical device or other entity that can be attached by a Driver service. See *Device Services* on page 188.
- *DriverLocator* – Assistant in locating bundles that provide a Driver service. See *Driver Locator Service* on page 201.
- *DriverSelector* – Assistant in selecting which Driver service is best suited to a Device service. See *The Driver Selector Service* on page 204.

Figure 16 show the classes and their relationships.

Figure 16   *Device Access Class Overview*

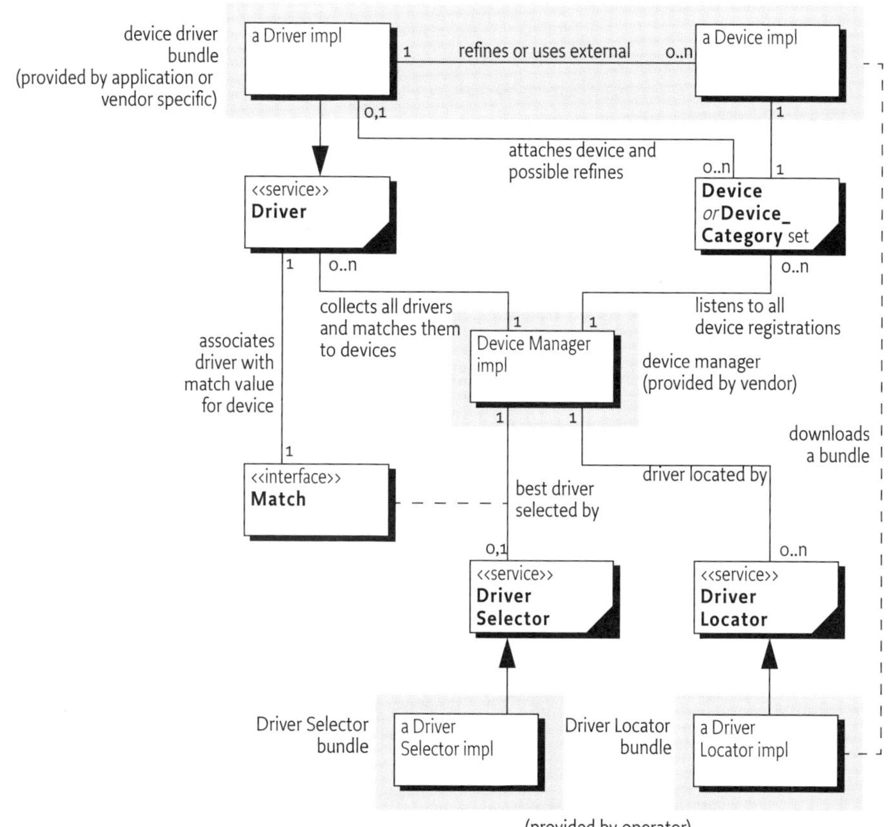

8.2 Device Services

A Device service represents some form of a device. It can represent a hardware device, but that is not a requirement. Device services differ widely: some represent individual physical devices and others represent complete networks. Several Device services can even simultaneously represent the same physical device at different levels of abstraction. For example:

- A USB network.
- A device attached on the USB network.
- The same device recognized as a USB to Ethernet bridge.
- A device discovered on the Ethernet using Salutation.
- The same device recognized as a simple printer.
- The same printer refined to a PostScript printer.

A device can also be represented in different ways. For example, a USB mouse can be considered as:

- A USB device which delivers information over the USB bus.
- A mouse device which delivers x and y coordinates and information about the state of its buttons.

Each representation has specific implications:

- That a particular device is a mouse is irrelevant to an application which provides management of USB devices.
- That a mouse is attached to a USB bus or a serial port would be inconsequential to applications that respond to mouse-like input.

Device services must belong to a defined *device category*, or else they can implement a generic service which models a particular device, independent of its underlying technology. Examples of this type of implementation could be a Sensor or Actuator service.

A device category specifies the methods to communicate with a Device service and enables interoperability between bundles that are based on the same underlying technology. Generic Device services will allow interoperability between bundles that are not coupled to specific device technologies.

For example, a device category is required for the USB, so that Driver bundles can be written that communicate to the devices that are attached to the USB. If a printer is attached, it should also be available as a generic Printer service defined in a Printer service specification, indistinguishable from such a Printer service that is attached to a parallel port. Generic categories, such as a Printer service, should also be described in a Device Category.

It is expected that most Device service objects will actually represent a physical device in some form, but that is not a requirement of this specification. A Device service is represented as a normal service in the OSGi Framework and all coordination and activities are performed upon Framework services. This specification does not limit a bundle developer from using Framework mechanisms for services that are not related to physical devices.

8.2.1 Device Service Registration

A Device service is defined as a normal service registered with the Framework that either:

- Registers a service object under the interface org.osgi.service.Device with the Framework, or
- Sets the DEVICE_CATEGORY property in the registration. The value of DEVICE_CATEGORY is an array of String objects of all the device categories that the device belongs to. These strings are defined in the associated device category.

If this document mentions a Device service, it is meant to refer to services registered with the name org.osgi.service.device.Device *or* services registered with the DEVICE_CATEGORY property set.

When a Device service is registered, additional properties may be set that describe the device to the device manager and potentially to the end users. The following properties have their semantics defined in this specification:

- DEVICE_CATEGORY – A marker property indicating that this service must be regarded as a Device service by the device manager. Its value is of type String[], and its meaning is defined in the associated device category specification.
- DEVICE_DESCRIPTION – Describes the device to an end user. Its value is of type String.
- DEVICE_SERIAL – A unique serial number for this device. If the device hardware contains a serial number, the driver bundle is encouraged to specify it as this property. Different Device services representing the same physical hardware at different abstraction levels should set the same DEVICE_SERIAL thus simplifying identification. Its value is of type String.
- service.pid – Service Persistent ID (PID), defined in org.osgi.framework.Constants. Device services should set this property. It must be unique among all registered services. Even different abstraction levels of the same device must use different PIDs. The service PIDs must be reproducible, so that every time the same hardware is plugged in, the same PIDs are used.

8.2.2	## Device Service Attachment

When a Device service is registered with the Framework, the device manager is responsible for finding a suitable Driver service and instructing it to attach to the newly registered Device service. The Device service itself is passive: it only registers a Device service with the Framework and then waits until it gets called.

The actual communication with the underlying physical device is not defined in the Device interface because it differs significantly between different types of devices. The Driver service is responsible for attaching the device in a device type-specific manner. The rules and interfaces for this process must be defined in the appropriate device category.

If the device manager is unable to find a suitable Driver service, the Device service remains unattached. In that case, if the service object implements the Device interface, it must receive a call to the noDriverFound() method. The Device service can wait until a new driver is installed, or it can unregister and attempt to register again with different properties that describe a more generic device or try a different configuration.

A Device service is not used by any other bundle according to the Framework; it is called an *idle Device service*.

8.2.2.1　**Device Service Unregistration**

When a Device service is unregistered, no immediate action is required by the device manager. The normal service unregistering events, provided by the Framework, take care of propagating the unregistration information to affected drivers. Drivers must take the appropriate action to release this Device service and perform any necessary cleanup as described in their device category specification.

The device manager may, however, take a device unregistration as an indication that driver bundles may have become idle and are thus eligible for removal. It is therefore important for Device services to unregister their service object when the underlying entity becomes unavailable.

8.3　Device Category Specifications

A device category specifies the rules and interfaces needed for the communication between a Device service and a Driver service. Only Device services and Driver services of the same device category can communicate and cooperate.

The Device Access service specification is limited to the attachment of Device services by Driver services, and does *not* enumerate different device categories.

Other specifications must specify a number of device categories before this specification can be made operational. Without a set of defined device categories, no interoperability can be achieved.

Device categories are related to a specific device technology, such as USB, IEEE 1394, JINI, UPnP, Salutation, CEBus, Lonworks, and others. The purpose of a device category specification is to make all Device services of that category conform to an agreed interface, so that, for example, a USB Driver service of vendor A can control Device services from vendor B attached to a USB bus.

This specification is limited to defining the guidelines for device category definitions only. Device categories may be defined by the OSGi or by external specification bodies – for example, when these bodies are associated with a specific device technology.

8.3.1 Device Category Guidelines

A device category definition comprises the following elements:

- An interface that all devices belonging to this category must implement. This interface should lay out the rules of how to communicate with the underlying device. The specification body may define its own device interfaces (or classes) or leverage existing ones. For example, a serial port device category could use the javax.comm.SerialPort interface which is defined in [20] *Java Communications API*.
  When registering a device belonging to this category with the Framework, the interface or class name for this category must be included in the registration.
- A set of service registration properties, their data types, and semantics, each of which must be declared as either MANDATORY or OPTIONAL for this device category.
- A range of match values specific to this device category. Matching is explained later in *The Device Attachment Algorithm* on page 207.

8.3.2 Sample Device Category Specification

The following is a partial example of a fictitious device category:

```
public interface com.acme.widget.WidgetDevice {
    int MATCH_SERIAL            = 10;
    int MATCH_VERSION           =  8;
    int MATCH_MODEL             =  6;
    int MATCH_MAKE              =  4;
```

```
int MATCH_CLASS                        = 2;
void sendPacket( byte [] data );
byte [] receivePacket( long timeout );
```
}

Devices in this category must implement the interface com.acme.widget.WidgetDevice to receive attachments from Driver services in this category.

Device properties for this fictitious category are defined in table Table 11.

Property name	M/O	Type	Value
DEVICE_CATEGORY	M	String[]	{"Widget"}
com.acme.class	M	String	A class description of this device. For example "audio", "video", "serial", etc. An actual device category specification should contain an exhaustive list and define a process to add new classes.
com.acme.model	M	String	A definition of the model. This is usually vendor specific. For example "Mouse".
com.acme.manufacturer	M	String	Manufacturer of this device, for example "ACME Widget Division".
com.acme.revision	O	String	Revision number. For example, "42".
com.acme.serial	O	String	A serial number. For example "SN6751293-12-2112/A".

Table 11　　　Example Device Category Properties, M=Mandatory, O=Optional

8.3.3　　Match Example

Driver services and Device services are connected via a matching process that is explained in *The Device Attachment Algorithm* on page 207. The Driver service plays a pivotal role in this matching process. It must inspect the Device service (from its ServiceReference object) that has just been registered and decide if it potentially could cooperate with this Device service.

It must be able to answer a value indicating the quality of the match. The scale of this match value must be defined in the device category so as to allow Driver services to match on a fair basis. The scale must start at least at 1 and go upwards.

Driver services for this sample device category must return one of the match codes defined in the com.acme.widget.WidgetDevice interface or Device.MATCH_NONE if the Device service is not recognized. The device category must define the exact rules for the match codes in the device category specification. In this example, a small range from 2 to 10 (MATCH_NONE is 0) is defined for WidgetDevice devices. They are named in the WidgetDevice interface for convenience and have the following semantics.

Match name	Value	Description
MATCH_SERIAL	10	An exact match, including the serial number.
MATCH_VERSION	8	Matches the right class, make model, and version.
MATCH_MODEL	6	Matches the right class and make model.
MATCH_MAKE	4	Matches the make.
MATCH_CLASS	2	Only matches the class.

Table 12   *Sample Device Category Match Scale*

A Driver service should use the constants to return when it decides how close the Device service matches its suitability. For example, if it matches the exact serial number, it should return MATCH_SERIAL.

8.4   Driver Services

A Driver service is responsible for attaching to suitable Device services under control of the device manager. Before it can attach a Device service, however, it must compete with other Driver services for control.

If a Driver service wins the competition, it must attach the device in a device category-specific way. After that, it can perform its intended functionality. This functionality is not defined here nor in the device category; this specification only describes the behavior of the Device service, not how the Driver service uses this to implement its intended functionality. A Driver service may register one or more new Device services of another device category or a generic service which models a more refined form of the device.

Both refined Device services as well as generic services should be defined in a Device Category. See *Device Category Specifications* on page 190.

8.4.1 Driver Bundles

A Driver service is, like *all* services, implemented in a bundle and is recognized by the device manager by registering one or more Driver service objects with the Framework.

Such bundles containing one or more Driver services are called *driver bundles.* The device manager must be aware of the fact that the cardinality of the relationship between bundles and Driver services is 1:1...n.

A driver bundle must register *at least* 1 Driver service in its Bundle-Activator.start implementation.

8.4.2 Driver Taxonomy

Device Drivers may belong to one of the following categories:

- Base Drivers (Discovery, Pure Discovery and Normal)
- Refining Drivers
- Network Drivers
- Composite Drivers
- Referring Drivers
- Bridging Drivers
- Multiplexing Drivers
- Pure Consuming Drivers

This list is not definitive, and a Driver service is not required to fit into one of these categories. The purpose of this taxonomy is to show the different topologies that have been considered for the Device Access service specification.

Figure 17   *Legend for Device Driver Services Taxonomy*

8.4.2.1 Base Drivers

The first category of device drivers are called *base drivers* because they provide the lowest-level representation of a physical device. The distinguishing factor is that they are not registered as a Driver service because they do not have to compete for access to their underlying technology.

Figure 18   *Base Driver Types*

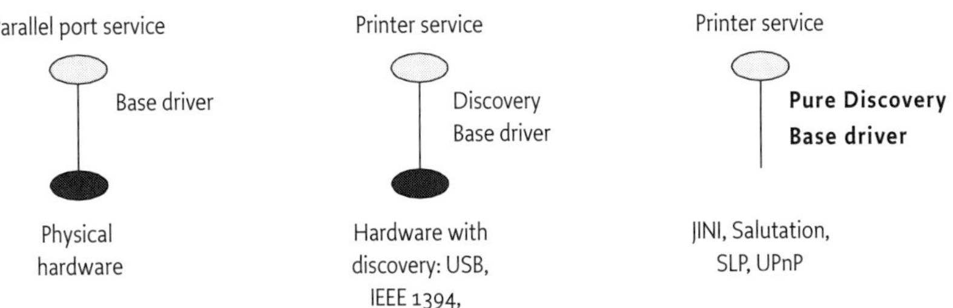

Base drivers discover physical devices using code not specified here (for example, through notifications from a device driver in native code) and then register corresponding Device services.

When the hardware supports a discovery mechanism and reports a physical device, a Device service is then registered. Drivers supporting a discovery mechanism are called *discovery base drivers.*

An example of a discovery base driver would be a USB driver. Discovered USB devices are registered with the Framework as a generic USB Device service. The USB specification, see [21] *USB Specification,* defines a tightly integrated discovery method. Further, devices are individually addressed and there is no provision for broadcasting a message to all devices attached to the USB bus. Therefore, there is no reason to expose the USB network itself; instead, a discovery base driver can register the individual devices as they are discovered.

Not all technologies support a discovery mechanism. For example, most serial devices do not support detection, and it is often not even possible to detect whether a device is attached to a serial port. Although a [22] *Plug and Play External COM Device Specification v. 1.0* exists, very few devices support it.

Although each driver bundle should perform discovery on its own, a driver for a non-discoverable serial port requires external help – either through a user interface or by allowing the Configuration Manager to configure it.

It is possible for the driver bundle to combine automatic discovery of Plug and Play-compliant devices with manual configuration when non-compliant devices are plugged in.

8.4.2.2   Refining Drivers

The second category of device drivers are called *refining drivers*. Refining drivers provide a refined view of a physical device that is already represented by another Device service registered with the Framework. Refining drivers register a Driver service with the Framework. This Driver service is used by the device manager to attach the refining driver to a less refined Device service that is registered as a result of events within the Framework itself.

Figure 19   *Refining Driver Diagram*

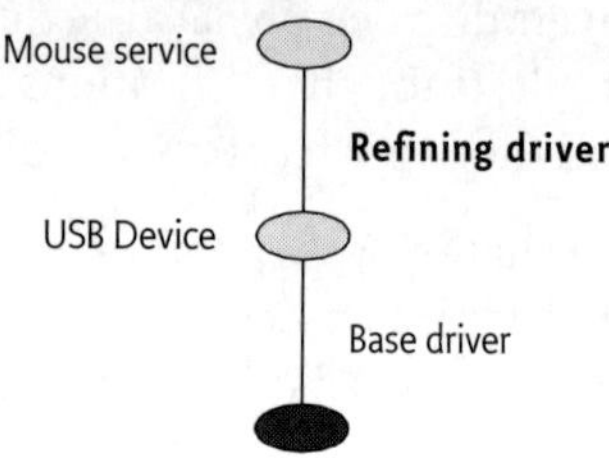

An example of a refining driver would be a mouse driver which is attached to the generic USB Device service representing a physical mouse. It then registers a new Device service which represents it as a Mouse service, defined elsewhere.

The majority of drivers fall into the refining driver type.

8.4.2.3   Network Drivers

An Internet Protocol (IP) capable network such as Ethernet supports individually addressable devices and allows broadcasts, but does not define an intrinsic discovery protocol. In this case, the entire network should be exposed as a single Device service.

Figure 20   *Network Driver diagram*

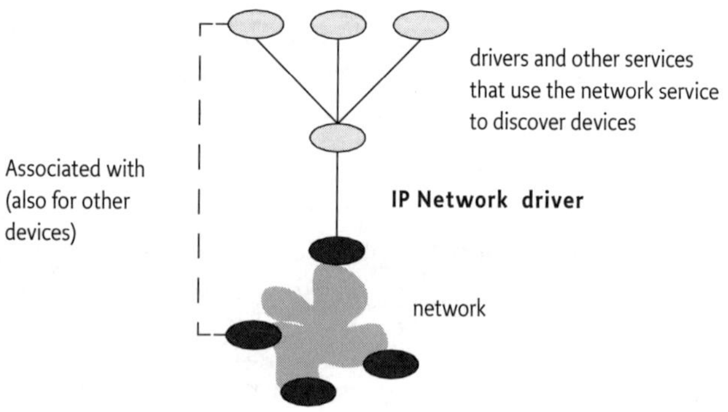

8.4.2.4   Composite Drivers

Complex devices can often be decomposed into several parts. Drivers that attach to a single service and then register multiple Device services are called *composite drivers*. For example, a USB speaker containing software accessible buttons can be registered by its driver as two separate Device services: an Audio Device service and a Button Device service.

Figure 21   *Composite Driver structure*

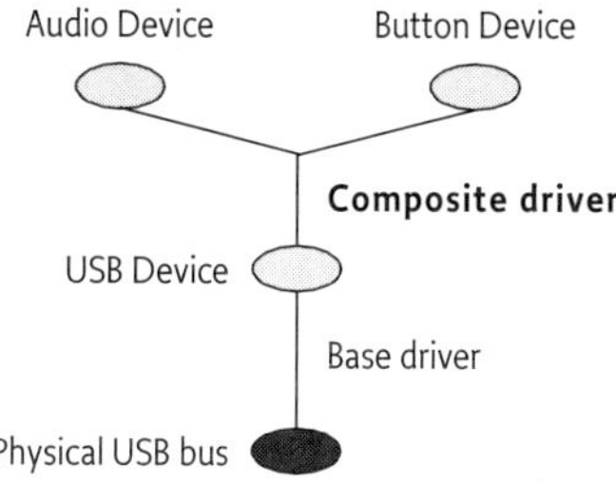

This approach can greatly reduce the number of interfaces needed, as well as enhance reusability.

8.4.2.5   Referring Drivers

A referring driver is actually not a driver in the sense that it controls Device services. Instead, it acts as an intermediary to help locate the correct driver bundle. This process is explained in detail in *The Device Attachment Algorithm* on page 207.

A referring driver implements the call to the attach method to inspect the Device service, and decides which Driver bundle would be able to attach to the device. This process can actually involve connecting to the physical device and communicating with it. The attach method then returns a String object that indicates the DRIVER_ID of another driver bundle. This process is called a referral.

For example, a vendor ACME can implement one driver bundle that specializes in recognizing all of the devices the vendor produces. The referring driver bundle does not contain code to control the device – it contains only sufficient logic to recognize the assortment of devices. This referring driver can be small, yet can still identify a large product line. This approach can drastically reduce the amount of downloading and matching needed to find the correct driver bundle.

8.4.2.6   Bridging Drivers

A bridging driver registers a Device service from one device category but attaches it to a Device service from another device category.

Figure 22   *Bridging Driver Structure*

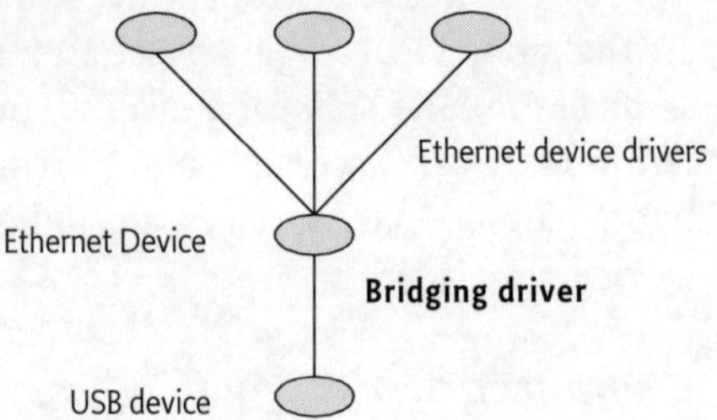

For example, USB to Ethernet bridges exist that allow connection to an Ethernet network through a USB device. In this case, the top level of the USB part of the Device service stack would be an Ethernet Device service. But the same Ethernet Device service can also be the bottom layer of an Ethernet layer of the Device service stack. A few layers up, a bridge could connect into yet another network.

The stacking depth of Device services has no limit, and the same drivers could in fact appear at different levels in the same Device service stack. The graph of drivers-to-Device services roughly mirrors the hardware connections.

8.4.2.7   Multiplexing Drivers

A *multiplexing driver* attaches a number of Device services and aggregates them in a new Device service.

Figure 23   *Multiplexing Driver Structure*

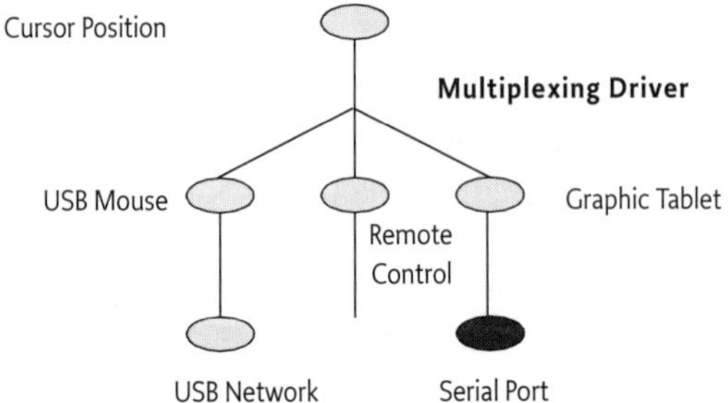

For example, a system has a mouse on USB, a graphic tablet on a serial port, and a remote control facility. Each of these would be registered as a service with the Framework. A multiplexing driver can attach all three, and can merge the different positions in a central Cursor Position service.

8.4.2.8   Pure Consuming Drivers

A *pure consuming driver* bundle will attach to devices without registering a refined version.

Figure 24   *Pure Consuming Driver Structure*

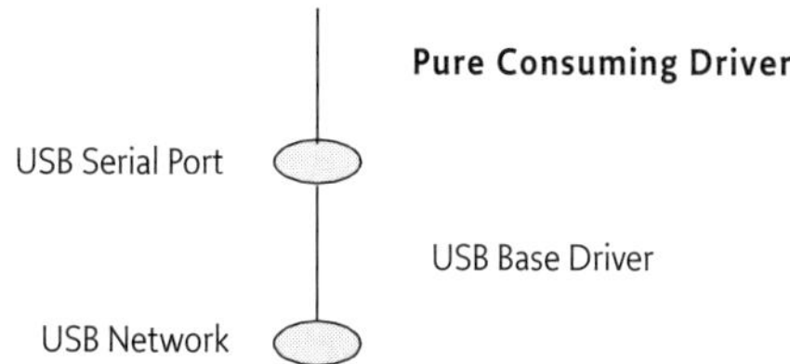

For example, one driver bundle could decide to handle all serial ports through javax.comm instead of registering them as services. When a USB serial port is plugged in, one or more Driver services are attached, resulting in a Device service stack with a Serial Port Device service. A pure consuming driver may then attach to the Serial Port Device service and register a new serial port with the javax.comm.* registry instead of the Framework service registry. This registration effectively transfers the device from the OSGi environment into another environment.

8.4.2.9   **Other Driver Types**

It should be noted that any bundle installed in the OSGi environment may get and use a Device service without having to register a Driver service.

The following functionality is offered to those bundles that do register a Driver service and conform to the this specification:

- The bundles can be installed and uninstalled on demand.
- Attachment to the Device service is only initiated after the winning the competition with other drivers.

8.4.3   **Driver Service Registration**

Drivers are recognized by registering a Driver service with the Framework. This event makes the device manager aware of the existence of the Driver service. A Driver service registration must have a DRIVER_ID property whose value is a String object, uniquely identifying the driver to the device manager. The device manager must use the DRIVER_ID to prevent the installation of duplicate copies of the same driver bundle.

Therefore, this DRIVER_ID must:

- Depend only on the specific behavior of the driver, and thus be independent of unrelated aspects like its location or mechanism of downloading.
- Start with the reversed form of the domain name of the company that implements it: for example, com.acme.widget.1.1.

- Differ from the DRIVER_ID of drivers with different behavior. Thus, it must *also* be different for each revision of the same driver bundle so they may be distinguished.

When a new Driver service is registered, the Device Attachment Algorithm must be applied to each idle Device service. This requirement gives the new Driver service a chance to compete with other Driver services for attaching to idle devices. The techniques outlined in *Optimizations* on page 211 can provide significant shortcuts for this situation.

As a result, the Driver service object can receive match and attach requests before the method which registered the service has returned.

This specification does not define any method for new Driver services to *steal* already attached devices. Once a Device service is attached by a Driver service, it can only be released by the Driver service itself.

8.4.4   Driver Service Unregistration

When a Driver service is unregistered, it must release all Device services to which it is attached. Thus *all* its attached Device services become idle. The device manager must gather all of these idle Device services and try to re-attach them. This condition gives other Driver services a chance to take over the refinement of devices after the unregistering driver. The techniques outlined in *Optimizations* on page 211 can provide significant shortcuts for this situation.

A Driver service that is installed by the device manager must remain registered as long as the driver bundle is active. Therefore, a Driver service should only be unregistered if the driver bundle is stopping, an occurrence which may precede being uninstalled or updated. Driver services should thus not unregister in an attempt to minimize resource consumption. Such optimizations can easily introduce race conditions with the device manager.

8.4.5   Driver Service Methods

The Driver interface consists of the following methods:

- match(ServiceReference) – This method is called by the device manager to find out how well this Driver service matches the Device service as indicated by the ServiceReference argument. The value returned here is specific for a device category. If this Device service is of another device category, the value Device.MATCH_NONE must be returned. Higher values indicate a better match. For the exact matching algorithm, see *The Device Attachment Algorithm* on page 207.

Driver match values and referrals must be deterministic in that repeated calls for the same Device service must return the same results so that results can be cached by the device manager.

- attach(ServiceReference) – If the device manager decides that a Driver service should be attached to a Device service, it must call this method on the Driver service object. Once this method is called, the Device service is regarded to be attached to that Driver service, and no other Driver service must be called to attach to the Device service. The Device service must remain *owned* by the Driver service until the Driver bundle is stopped. No unattach method exists.

  The attach method should return null when the Device service is correctly attached. A referring driver (see *Referring Drivers* on page 197) can return a String object that specifies the DRIVER_ID of a driver that can handle this Device service. In this case, the Device service is not attached and the device manager must attempt to install a Driver service with the same DRIVER_ID via a Driver Locator service.

  The attach method must be deterministic as described in the previous method.

8.4.6   Idle Driver Bundles

An idle Driver bundle is a bundle with a registered Driver service and is not attached to any Device service. Idle Driver bundles are consuming resources in the OSGi environment. The device manager should uninstall bundles that it has installed which are idle.

8.5   Driver Locator Service

The device manager must automatically install Driver bundles, which are obtained from Driver Locator services, when new Device services are registered.

A Driver Locator service encapsulates the knowledge of how to fetch the Driver bundles needed for a specific Device service. This selection is made on the properties that are registered with a device: for example, DEVICE_CATEGORY and any other properties registered with the Device service registration.

The purpose of the Driver Locator service is to separate the mechanism from the policy. The decision to install a new bundle is made by the device manager (the mechanism), but a Driver Locator service decides which bundle to install and from where the bundle is downloaded the policy).

Installing bundles has many consequences for the security of the system, and this process is also sensitive to network setup and other configuration details. Using Driver Locator services allows the service gateway operator to choose a strategy that best fits its needs.

Driver services are identified by the DRIVER_ID property. Driver Locator services use this particular ID to identify the bundles that can be installed. Driver ID properties have uniqueness requirements as specified in *Device Service Registration* on page 189. This uniqueness allows the device manager to maintain a list of Driver services and prevent unnecessary installs.

An OSGi environment can have several different Driver Locator services installed. The device manager must consult all of them and use the combined result set, after pruning duplicates based on the DRIVER_ID values.

8.5.1   The DriverLocator Interface

The DriverLocator interface allows suitable driver bundles to be located, downloaded, and installed on demand, even when completely unknown devices are detected.

It has the following methods:

- findDrivers(Dictionary) – This method returns an array of driver IDs that potentially match a service that is described by the properties in the Dictionary object. A driver ID is the String object that is registered by a Driver service under the DRIVER_ID property.
- loadDriver(String) – This method returns an InputStream object that can be used to download the bundle containing the Driver service as specified by the driver ID argument. If the Driver Locator service cannot download such a bundle, it should return null. Once this bundle is downloaded and installed in the Framework, it must register a Driver service with the DRIVER_ID property set to the value of the String argument.

8.5.2   A Driver Example

The following example shows a very minimal Driver service implementation. It consists of two classes. The first class is SerialWidget. This class tracks a single WidgetDevice from *Sample Device Category Specification* on page 191. It registers a javax.comm.SerialPort service which is a general serial port specification that could also be implemented from other device categories like USB, a COM port, etc.. It is created when the SerialWidgetDriver object is requested to attach a WidgetDevice by the device manager. It registers a new javax.comm.SerialPort service in its constructor.

The org.osgi.util.tracker.ServiceTracker is extended to handle the Framework events that are needed to simplify tracking this service. The removedService method of this class is overridden to unregister the SerialPort when the underlying WidgetDevice is unregistered.

```
package com.acme.widget;
import org.osgi.service.device.*;
import org.osgi.framework.*;
import org.osgi.util.tracker.*;

class SerialWidget extends ServiceTracker
    implements javax.comm.SerialPort,
       org.osgi.service.device.Constants {
       ServiceRegistration     registration;

    SerialWidget( BundleContext c, ServiceReference r
) {
        super( c, r, null );
        open();
    }

    public Object addingService( ServiceReference ref
) {
        WidgetDevice dev = (WidgetDevice)
          context.getService( ref );
        registration = context.registerService(
            javax.comm.SerialPort.class.getName(),
            this,
            null
          );
        return(dev);
    }

    public void removedService( ServiceReference ref,
      Object service
    ) {
        registration.unregister();
        context.ungetService(ref);
    }
    ... methods for javax.comm.SerialPort that are
    ... converted to underlying WidgetDevice
}
```

A SerialWidgetDriver object is registered with the Framework in the Bundle Activator start method under the Driver interface. The device manager must call the match method for each idle Device service that is registered. If it is chosen by the device manager to control this Device service, a new SerialWidget is created that offers serial port functionality to other bundles.

```java
public class SerialWidgetDriver implements Driver {
    BundleContext          context;

    String     spec =
      "(&"
    +"  (objectclass=com.acme.widget.WidgetDevice)"
    +"   (DEVICE_CATEGORY=WidgetDevice)"
    +"   (com.acme.class=Serial)"
    + ")";

    Filter    filter;

    SerialWidgetDriver( BundleContext context )
      throws Exception {
      this.context = context;
      filter = context.createFilter(spec);
    }

    public int match( ServiceReference d ) {
      if ( filter.match( d ) )
        return WidgetDevice.MATCH_CLASS
      else
        return Device.MATCH_NONE;
    }

    public synchronized String attach(
ServiceReference r ) {
      new SerialWidget( context, r );
    }
}
```

8.6 The Driver Selector Service

The purpose of the Driver Selector service is to customize the selection of the best Driver service from a set of suitable Driver bundles. The device manager has a default algorithm as described in *The Device Attachment Algorithm* on page 207. When this algorithm is not sufficient and requires customizing by the operator, a bundle providing a Driver Selector service can be installed in the Framework. This service must be used by the device manager as the final arbiter when select the best match for a Device service.

The Driver Selector service is a singleton; only one such service is recognized by the device manager. The Framework method BundleContext.getServiceReference must be used to obtain a Driver Selector service. In the erroneous case that multiple Driver Selector services are registered, the service.ranking property will thus define which actual service is used.

The Driver Selector service implements the DriverSelector interface.

A device manager implementation must invoke the method select(ServiceReference, Match[]). This method receives a Service Reference to the Device service and an array of Match objects. Each Match object contains a link to the ServiceReference object of a Driver service and the result of the match value returned from a previous call to Driver.match. The Driver Selector service should inspect the array of Match objects and use some means to decide which Driver service is best suited. The index of the best match should be returned. If none of the Match objects describe a possible Driver service, the implementation must return DriverSelector.SELECT_NONE (-1).

8.7   Device Manager

Device Access is controlled by the device manager in the background. The device manager is responsible for initiating all actions in response to the registration, modification, and unregistration of Device services and Driver services, using Driver Locator services and a Driver Selector service as helpers.

The device manager detects the registration of Device services and coordinates their attachment with a suitable Driver service. Potential Driver services do not have to be active in the Framework to be eligible. The device manager must use Driver Locator services to find bundles that might be suitable for the detected Device service and that are not currently installed. This selection is done via a DRIVER_ID property that is unique for each Driver service.

The device manager must install and start these bundles with the help of a Driver Locator service. This activity must result in the registration of one or more Driver services. All available Driver services, installed by the device manager and also others, then participate in a bidding process. The Driver service can inspect the Device service through its ServiceReference object to find out how well this Driver service matches the Device service.

If a Driver Selector service is available in the Framework service registry, it is used to decide which of the eligible Driver services is the best match.

If no Driver Selector service is available, the highest bidder must win, with tie breaks defined on the service.ranking and service.id properties. The selected Driver service is then asked to attach the Device service.

If no Driver service is suitable, the Device service remains idle. When new Driver bundles are installed, these idle Device services must be reattached.

The device manager must reattach a Device service if at a later time a Driver service is unregistered due to an uninstallation or update. At the same time, however, it should prevent superfluous and non optimal reattachments. The device manager should also garbage collect driver bundles it installed which are no longer used.

The device manager is a singleton. There must only be one device manager registered with a Framework, and it has no public interface.

8.7.1 Device Manager Startup

To prevent race conditions during Framework startup, the device manager must monitor the state of Device services and Driver services immediately when it is started. The device manager must not, however, begin attaching Device services until the Framework has been fully started, to prevent superfluous or non-optimal attachments.

The Framework has completed starting when the FrameworkEvent.STARTED event has been published. Publication of that event indicates that Framework has finished all its initialization and all bundles are started. If the device manager is started after the Framework has been initialized, it should detect the state of the Framework by examining the state of the system bundle.

8.7.2   The Device Attachment Algorithm

A key responsibility of the device manager is to attach refining drivers to idle devices. The following diagram illustrates the device attachment algorithm.

Figure 25   *Device Attachment Algorithm*

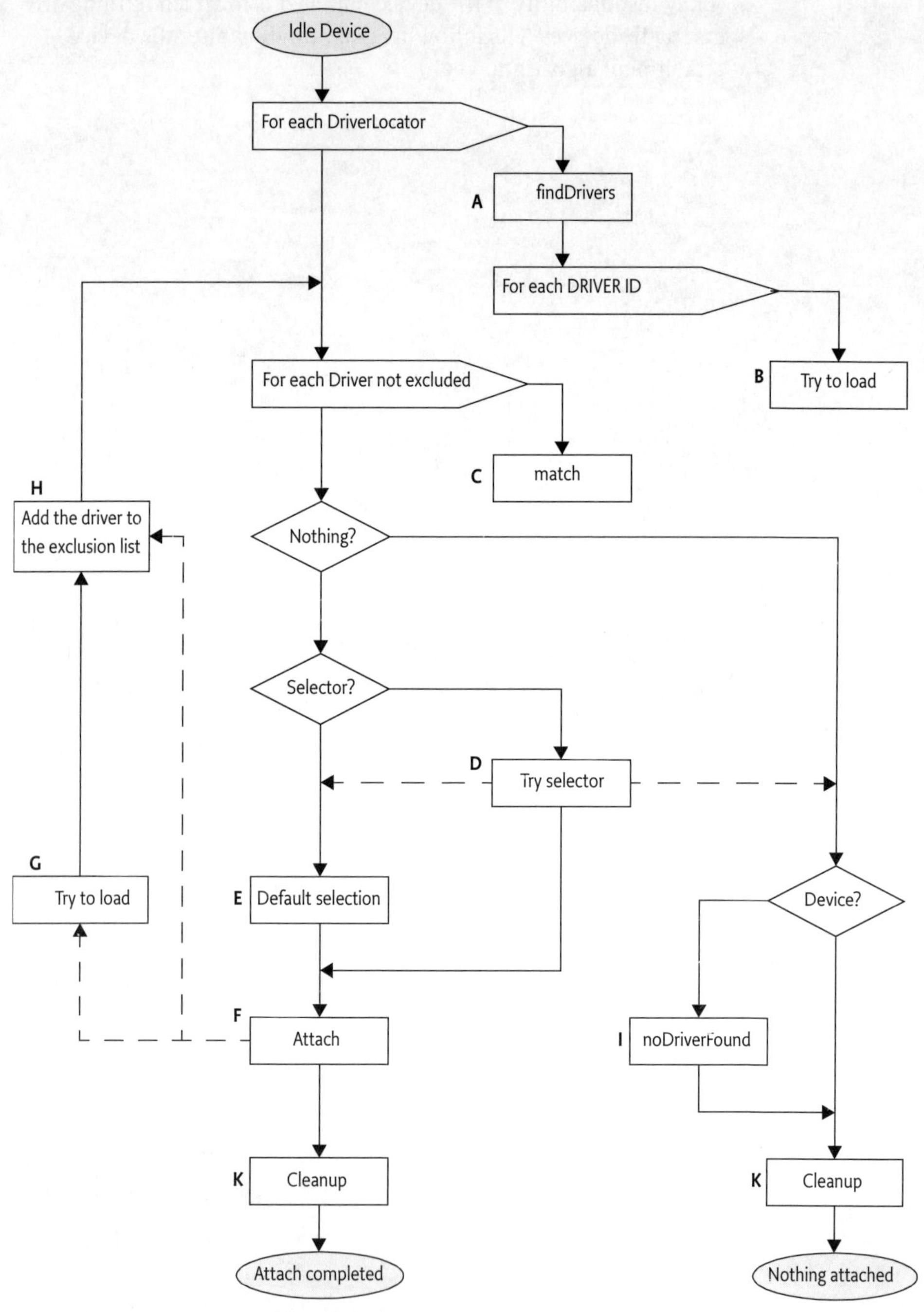

8.7.3

Step	Legend Description
A	DriverLocator.findDrivers is called for each registered Driver Locator service, passing the properties of the newly detected Device service. Each method call returns zero or more DRIVER_ID values (identifiers of particular driver bundles). If the findDrivers method throws an exception, it is ignored, and processing continues with the next Driver Locator service. See *Optimizations* on page 211 for further guidance on handling exceptions.
B	For each found DRIVER_ID that does not correspond to an already registered Driver service, the device manager calls DriverLocator.loadDriver to return an InputStream containing the driver bundle. Each call to loadDriver is directed to one of the Driver Locator services that mentioned the DRIVER_ID in Step A. If the loadDriver method fails, the other Driver Locator objects are tried. If they all fail, the driver bundle is ignored. If this method succeeds, the device manager installs and starts the driver bundle. Driver bundles must register their Driver services synchronously during bundle activation.
C	For each Driver service, except those on the exclusion list, call its Driver.match method passing the ServiceReference object to the Device service. Collect all successful matches – that is, those whose return values are greater than Device.MATCH_NONE – in a list of active matches. A match call that throws an exception is considered unsuccessful and is not added to the list.

Table 13   *Driver attachment algorithm*

Step	Description
D	If there is a Driver Selector service, the device manager calls the DriverSelector.select method, passing the array of active Match objects.
	If the Driver Selector service returns the index of one of the Match objects from the array, its associated Driver service is selected for attaching the Device service. If the Driver Selector service returns DriverSelector.SELECT_NONE, no Driver service must be considered for attaching the Device service.
	If the Driver Selector service throws an exception or returns an invalid result, the default selection algorithm is used.
	Only one Driver Selector service is used even if there is more than one registered in the Framework. See *The Driver Selector Service* on page 204.
E	The winner is the one with the highest match value. Tie breakers are respectively:
	• Highest service.ranking property. • Lowest service.id property.
F	The selected Driver service's attach method is called. If the attach method returns null, the Device service has been successfully attached. If the attach method returns a String object, it is interpreted as a referral to another Driver service and processing continues at G. See *Referring Drivers* on page 197.
	If an exception is thrown, the Driver service has failed and the algorithm proceeds to try another Driver service after excluding this one from further consideration at Step H.
G	The device manager attempts to load the referred driver bundle in a manner similar to Step B, except that it is unknown which Driver Locator service to use. Therefore, the loadDriver method must be called on each Driver Locator service until one succeeds (or they all fail). If one succeeds, the device manager installs and starts the driver bundle. The driver bundle must register a Driver service during its activation which must be added to the list of Driver services in this algorithm.

Table 13   *Driver attachment algorithm*

Step	Description
H	The referring driver bundle is added to the exclusion list. Because each new referral adds an entry to the exclusion list, which in turn disqualifies another driver from further matching, the algorithm cannot loop indefinitely. This list is maintained for the duration of this algorithm. The next time a new Device service is processed, the exclusion list starts out empty.
I	If no Driver service attached the Device service, the Device service is checked to see whether it implements the Device interface. If so, the noDriverFound method is called. Note that this may cause the Device service to unregister and possibly a new Device service (or services) to be registered in its place. Each new Device service registration must restart the algorithm from the beginning.
K	Whether an attachment was successful or not, the algorithm may have installed a number of driver bundles. The device manager should remove any idle driver bundles that it installed.

Table 13   *Driver attachment algorithm*

8.7.4 Optimizations

Optimizations are explicitly allowed and even recommended for an implementation of a device manager. Implementations may use the following assumptions:

- Driver match values and referrals must be deterministic in that repeated calls for the same Device service must return the same results.
- The device manager may cache match values and referrals. Therefore, optimizations in the device attachment algorithm based on this assumption are allowed.
- The device manager may delay loading a driver bundle until it is needed. For example, a delay could occur when that DRIVER_ID's match values are cached.
- The results of calls to DriverLocator and DriverSelector methods are not required to be deterministic, and must not be cached by the device manager.
- Thrown exceptions must not be cached. Exceptions are considered transient failures, and the device manager must always retry a method call even if it has thrown an exception on a previous invocation with the same arguments.

8.7.5   ## Driver Bundle Reclamation

The device manager may remove driver bundles it has installed at any time, provided that all the Driver services in that bundle are idle. This recommended practice prevents unused driver bundles from accumulating over time. Removing driver bundles too soon, however, may cause unnecessary installs and associated delays when driver bundles are needed again.

If a device manager implements driver bundle reclamation, the specified matching algorithm is not guaranteed to terminate unless the device manager takes reclamation into account.

For example, a new Device service triggers the attachment algorithm. A driver bundle recommended by a Driver Locator service is loaded. It does not match, so the Device service remains idle. The device manager is eager to reclaim space, and unloads the driver bundle. The disappearance of the Driver service causes the device manager to reattach idle devices. Not keeping record of its previous activities, it tries to reattach the same device, which closes the loop.

On systems where the device manager implements driver bundle reclamation, all refining drivers should be loaded through Driver Locator services. This recommendation is to prevent the device manager from erroneously uninstalling pre-installed driver bundles that cannot later be reinstalled when needed.

The device manager can be updated or restarted. It cannot, however, rely on previously stored information to determine which driver bundles were pre-installed and which were dynamically installed and thus are eligible for removal. The device manager may persistently store cachable information for optimization, but must be able to cold start without any persistent information and still be able to manage an existing connection state, satisfying all of the requirements in this specification.

8.7.6   ## Handling Driver Bundle Updates

It is not straightforward to determine whether a driver bundle is being updated when the UNREGISTER event for a Driver service is received. In order to facilitate this distinction, the device manager should wait a period of time after the unregistration for one of the following events:

- A BundleEvent.UNINSTALLED event for the driver bundle.
- A ServiceEvent.REGISTERED event for another Driver service registered by the driver bundle.

If the driver bundle is uninstalled, or if neither of the above events are received within the allotted time period, the driver is assumed to be inactive. The appropriate waiting period is implementation-dependent and will vary for different installations. As a general rule, it should be long enough to allow a driver to be stopped, updated, and restarted under normal conditions, and short enough not to cause unnecessary delays in reattaching devices. The actual time should be configurable.

8.7.7 Simultaneous Device Service and Driver Service Registration

The device attachment algorithm may cause driver bundles to be installed, which requires executing the device attachment algorithm recursively. In this case, the appearance of the new driver bundles should be queued until completion of the device attachment algorithm.

Only one device attachment algorithm may be in progress at any moment in time.

The following example sequence illustrates this process when a Driver service is registered:

- Collect the set of all idle devices.
- Apply the device attachment algorithm to each device in the set.
- If no Driver services were registered during the execution of the device attachment algorithm, processing terminates.
- Otherwise, restart this process.

8.8 Security

The device manager is the only privileged bundle in the Device Access specification and requires the org.osgi.AdminPermission to install and uninstall driver bundles.

The device manager itself should be free from any knowledge of policies and should not actively set bundle permissions. Rather, if permissions must be set, it is up to the management bundle to listen to synchronous bundle events and set the appropriate permissions.

Driver Locator services can trigger the download of any bundle, because they deliver the content of a bundle to the privileged device manager and could potentially insert a Trojan horse into the environment. Therefore, Driver Locator bundles need the ServicePermission[REGISTER] to register Driver Locator services, and the operator should exercise prudence in assigning this ServicePermission.

Bundles with Driver Selector services only require ServicePermission[REGISTER] to register the DriverSelector service. The Driver Selector service can play a crucial role in the selection of a suitable Driver service but it has no means to define a specific bundle itself.

8.9 Changes Since 1.0

- *Match Disambiguation* – In the Device Access 1.0 specification, if the matching process results in a tie, the device manager is free to choose any one of the highest bidders. The choice need not be consistent if the device manager is faced with the same situation again. This kind of randomness may surprise a user and also makes automated testing difficult. To avoid this situation, the selection now also involves the service.ranking and service.id properties. See *The Device Attachment Algorithm* on page 207 for more information.
- *Constants interface* – The Constants interface has been added. It contains constants for standard service property keys.
- *Driver Selector service* – A DriverSelector interface has been added. It allows complete customizing of the selection process. See *The Driver Selector Service* on page 204.
  The device manager by default eagerly attaches all the drivers it can without regard to whether the resulting refined devices are wanted or not. This eagerness could lead to installing and starting many unnecessary drivers. The Driver Selector service also provides a way to limit this eagerness.
- *Match Interface* – The Match interface has been added. It is used in the second argument to DriverSelector.select.
- *Tracking Driver Services* – The Device Access 1.0 specification prescribed that device managers should track the appearance of Device services. This requirement causes a problem if resident drivers are started in the *wrong* order. If a low-level Driver service is registered before a higher level driver bundle is started, the device manager could conclude that no refining driver is available, and leave the device unattached.
  This situation has been improved in two ways. First, the device manager must not start until the Framework has completed starting. Second, the device manager now must listen to Driver service registrations and unregistrations.
- *Free Format Devices* – The Device Access 1.0 specification defines Device services as services that implement the Device interface. This definition rules out existing interfaces that were not designed with the Device interface in mind: for example, javax.comm port interfaces. In this release, any service registered

with the DEVICE_CATEGORY property is defined to be a Device service. See *Device Services* on page 188 for more information.

8.10   org.osgi.service.device

The OSGi Device Access Package. Specification Version 1.1.

Bundles wishing to use this package must list the package in the Import-Package header of the bundle's manifest. For example:

```
Import-Package: org.osgi.service.device;
specification-version=1.1
```

Class Summary

Interfaces

Constants	This interface defines standard names for property keys associated with Device and Driver services.
Device	Interface for identifying device services.
Driver	A Driver service object must be registered by each Driver bundle wishing to attach to Device services provided by other drivers.
DriverLocator	A Driver Locator service can find and load device driver bundles given a property set.
DriverSelector	When the device manager detects a new Device service, it calls all registered Driver services to determine if anyone matches the Device service.
Match	Instances of Match are used in the select(ServiceReference, Match[]) method to identify Driver services matching a Device service.

8.10.1   Constants

public interface Constants

This interface defines standard names for property keys associated with Device and Driver services.

The values associated with these keys are of type java.lang.String, unless otherwise stated.

Since: 1.1

See Also: Device, Driver

8.10.1.1   **Fields**

public static final java.lang.String **DEVICE_CATEGORY**

Property (named "DEVICE_CATEGORY") containing a human readable description of the device categories implemented by a device. This property is of type String[]

Services registered with this property will be treated as devices and discovered by the device manager

public static final java.lang.String **DEVICE_DESCRIPTION**

Property (named "DEVICE_DESCRIPTION") containing a human readable string describing the actual hardware device.

public static final java.lang.String **DEVICE_SERIAL**

Property (named "DEVICE_SERIAL") specifying a device's serial number.

public static final java.lang.String **DRIVER_ID**

Property (named "DRIVER_ID") identifying a driver.

A DRIVER_ID should start with the reversed domain name of the company that implemented the driver (e.g., com.acme), and must meet the following requirements:

- It must be independent of the location from where it is obtained.
- It must be independent of the DriverLocator service that downloaded it.
- It must be unique.
- It must be different for different revisions of the same driver.

This property is mandatory, i.e., every Driver service must be registered with it.

8.10.2   **Device**

public interface Device

Interface for identifying device services.

A service must implement this interface or use the Constants.DEVICE_CATEGORY registration property to indicate that it is a device. Any services implementing this interface or registered with the DEVICE_CATEGORY property will be discovered by the device manager.

Device services implementing this interface give the device manager the opportunity to indicate to the device that no drivers were found that could (further) refine it. In this case, the device manager calls the noDriverFound() method on the Device object.

Specialized device implementations will extend this interface by adding methods appropriate to their device category to it.

See Also: Driver

8.10.2.1   **Fields**

public static final int **MATCH_NONE**

Return value from Driver.match(ServiceReference) indicating that the driver cannot refine the device presented to it by the device manager. The value is 0.

8.10.2.2   **Methods**

public void **noDriverFound**()

Indicates to this Device object that the device manager has failed to attach any drivers to it.

If this Device object can be configured differently, the driver that registered this Device object may unregister it and register a different Device service instead.

8.10.3   Driver

public interface Driver

A Driver service object must be registered by each Driver bundle wishing to attach to Device services provided by other drivers. For each newly discovered Device object, the device manager enters a bidding phase. The Driver object whose match(ServiceReference) method bids the highest for a particular Device object will be instructed by the device manager to attach to the Device object.

See Also: Device, DriverLocator

8.10.3.1   **Methods**

public java.lang.String **attach**(ServiceReference reference)
     throws java.lang.Exception

Attaches this Driver service to the Device service represented by the given ServiceReference object.

A return value of null indicates that this Driver service has successfully attached to the given Device service. If this Driver service is unable to attach to the given Device service, but knows of a more suitable Driver service, it must return the DRIVER_ID of that Driver service. This allows for the implementation of referring drivers whose only purpose is to refer to other drivers capable of handling a given Device service.

After having attached to the Device service, this driver may register the underlying device as a new service exposing driver-specific functionality.

This method is called by the device manager.

Parameters: reference - the ServiceReference object of the device to attach to

Returns: null if this Driver service has successfully attached to the given Device service, or the DRIVER_ID of a more suitable driver

Throws: java.lang.Exception - if the driver cannot attach to the given device and does not know of a more suitable driver

```
public int match(ServiceReference reference)
    throws java.lang.Exception
```

Checks whether this Driver service can be attached to the Device service represented by the given ServiceReference and returns a value indicating how well this driver can support the given Device service, or Device.MATCH_NONE if it cannot support the given Device service at all.

The return value must be one of the possible match values defined in the device category definition for the given Device service, or Device.MATCH_NONE if the category of the Device service is not recognized.

In order to make its decision, this Driver service may examine the properties associated with the given Device service, or may get the referenced service object (representing the actual physical device) to talk to it, as long as it ungets the service and returns the physical device to a normal state before this method returns.

A Driver service must always return the same match code whenever it is presented with the same Device service.

The match function is called by the device manager during the matching process.

Parameters: reference - the ServiceReference object of the device to match

Returns:	value indicating how well this driver can support the given Device service, or Device.MATCH_NONE if it cannot support the Device service at all
Throws:	java.lang.Exception - if this Driver service cannot examine the Device service

8.10.4 DriverLocator

public interface DriverLocator

A Driver Locator service can find and load device driver bundles given a property set. Each driver is represented by a unique DRIVER_ID.

Driver Locator services provide the mechanism for dynamically downloading new device driver bundles into an OSGi environment. They are supplied by providers and encapsulate all provider-specific details related to the location and acquisition of driver bundles.

See Also:	Driver

8.10.4.1 Methods

public java.lang.String[] **findDrivers**(java.util.Dictionary props)

Returns an array of DRIVER_ID strings of drivers capable of attaching to a device with the given properties.

The property keys in the specified Dictionary objects are case-insensitive.

Parameters:	props - the properties of the device for which a driver is sought
Returns:	array of driver DRIVER_ID strings of drivers capable of attaching to a Device service with the given properties, or null if this Driver Locator service does not know of any such drivers

public java.io.InputStream **loadDriver**(java.lang.String id)
   throws java.io.IOException

Get an InputStream from which the driver bundle providing a driver with the giving DRIVER_ID can be installed.

Parameters:	id - the DRIVER_ID of the driver that needs to be installed.
Returns:	a InputStream object from which the driver bundle can be installed
Throws:	java.io.IOException - the input stream for the bundle cannot be created

8.10.5 DriverSelector

public interface DriverSelector

When the device manager detects a new Device service, it calls all registered Driver services to determine if anyone matches the Device service. If at least one Driver service matches, the device manager must choose one. If there is a Driver Selector service registered with the Framework, the device manager will ask it to make the selection. If there is no Driver Selector service, or if it returns an invalid result, or throws an Exception, the device manager uses the default selection strategy.

Since: 1.1

8.10.5.1 Fields

public static final int **SELECT_NONE**

Return value from Driver Selector.select, if no Driver service should be attached to the Device service. The value is -1.

8.10.5.2 Methods

public int **select**(ServiceReference reference, Match[] matches)

Select one of the matching Driver services. The device manager calls this method if there is at least one driver bidding for a device. Only Driver services that have responded with nonzero (not MATCH_NONE)match values will be included in the list.

Parameters: reference - the ServiceReference object of the Device service.

matches - the array of all non-zero matches.

Returns: index into the array of Match objects, or SELECT_NONE if no Driver service should be attached

8.10.6 Match

public interface Match

Instances of Match are used in the select(ServiceReference, Match[]) method to identify Driver services matching a Device service.

Since: 1.1

See Also: DriverSelector

8.10.6.1 Methods

public ServiceReference **getDriver**()

Return the reference to a Driver service.

Returns: ServiceReference object to a Driver service.

public int **getMatchValue**()

Return the match value of this object.

Returns: the match value returned by this Driver service.

8.11 References

[20] *Java Communications API*
http://java.sun.com/products/javacomm

[21] *USB Specification*
http://www.usb.org/developers/data/usbspec.zip

[22] *Plug and Play External COM Device Specification v. 1.0*
http://www.microsoft.com/hwdev/download/respec/pnpcom.rtf

[23] *Universal Plug and Play*
http://www.upnp.org/resources.htm

[24] *Jini, Service Discovery and Usage*
http://www.jini.org/resources/

[25] *Salutation, Service Discovery Protocol*
http://www.salutation.org

9 Configuration Admin Service Specification

Version 1.0

9.1 Introduction

The Configuration Admin service is an important aspect of the deployment of an OSGi environment. It allows an operator to set the configuration information of deployed bundles.

Configuration is the process of defining the configuration data of bundles and assuring that those bundles receive that data when they are active in the OSGi environment.

Figure 26   *Configuration Admin Service Overview*

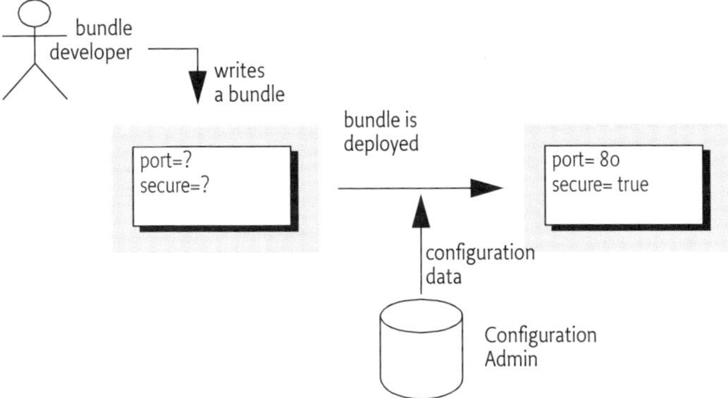

9.1.1 Essentials

The following requirements and patterns are related to the Configuration Admin service specification:

- *Local Configuration* – Must support bundles that have their own user interface to change their configurations.
- *Reflection* – Must be able to deduce the names and types of the needed configuration data.
- *Legacy* – Must support configuration data of existing entities (such as devices).
- *Object Oriented* – Must support the creation and deletion of instances of configuration information, so that a bundle can

create the appropriate number of services under control of the Configuration Admin service.

- *Embedded Devices* – Must be deployable on a wide range of platforms. This requirement means that the interface should not assume file storage on the platform. The choice to use file storage should be left to the implementation of the Configuration Admin service.
- *Remote versus Local Management* – Must allow for remotely managed OSGi environments, and must not assume that configuration is stored locally, nor should it assume that the Configuration Admin service is always done remotely; both implementation approaches should be viable.
- *Availability* – The OSGi environment is a dynamic environment that must run continuously (24/7/365). Configuration updates must happen dynamically and should not require restarting of the system or bundles.
- *Immediate Response* – Changes in the configuration should be reflected immediately.
- *Execution Environment* – Will not require more than the minimal execution environment.
- *Communications* – The Configuration Admin service should not assume "always-on" connectivity, so that the API is also applicable for mobile applications in cars, phones, or boats.
- *Extendability* – Expose the process of configuration to other bundles. This exposure should at least encompass initiating an update, removing certain configuration properties, adding properties, and modifying the value of properties potentially based on existing property or service values.
- *Complexity Trade-offs* – Bundles in need of configuration data should have a simple way of obtaining this data. Most bundles will have this need and the code to accept this data. Additionally, updates should be simple from the perspective of the receiver. Trade-offs in simplicity should be made at the expense of the bundle implementing the Configuration Admin service and in favor of bundles that need configuration information. The reason for this choice is that there will be many normal bundles and few Configuration Admin bundles.

9.1.2 Operation

This specification is based on the concept of a Configuration Admin service which manages the configuration of an OSGi environment. It maintains a database of Configuration objects, locally or remote. This service monitors the service registry and provides configuration information to services that are registered with a service.pid property, the Persistent IDentity (PID), and implement one of the following interfaces:

- *Managed Service* – A service registered with this interface receives its *configuration dictionary* from the database or null.
- *Managed Service Factory* – Services registered with this interface receive several configuration dictionaries when registered. The database contains zero or more configuration dictionaries for this service. Each configuration dictionary is given sequentially to this service.

The database can be manipulated by management bundles or bundles that configure themselves.

Third parties can provide a Configuration Plugin services. Such services participate in the configuration process. They can inspect the configuration dictionary and modify it before it reaches the target service.

9.1.3 Entities

- *Configuration information* – The information needed by a bundle before it can provide its intended functionality.
- *Configuration dictionary* – The configuration information when it is passed to the target service. It consists of a Dictionary object with a number of properties and identifiers.
- *Configuring Bundle* – The bundle that modifies the configuration information through the Configuration Admin service. This bundle is either a management bundle or the bundle for which the configuration information is intended.
- *Configuration Target* – The target (bundle or service) that will receive the configuration information. For services, there are two types of targets: ManagedServiceFactory or ManagedService objects.
- *Configuration Admin Service* – This service is responsible for supplying configuration target bundles with their configuration information. It maintains a database with configuration information, keyed on the service.pid of configuration target services. These objects receive their configuration dictionary or dictionaries when they are registered with the Framework. Configurations can be modified or extended using Configuration Plugin services before they reach the target bundle.
- *Managed Service* – A Managed Service represents a client of Configuration Admin service, and is thus a configuration target. Bundles should register a Managed Service to receive the configuration data from the Configuration Admin service. A Managed Service adds a unique service.pid service registration property as a primary key for the configuration information.
- *Managed Service Factory* – A Managed Service Factory can receive a number of configuration dictionaries from the Configuration Admin service, and is thus also a configuration target service. It

should register with a service.pid and receives zero or more configuration dictionaries; each dictionary has its own PID.

- *Configuration Object* – Implements the Configuration interface and contains the configuration dictionary for a Managed Service or a single instance for a Managed Service Factory. These objects are manipulated by configuring bundles.
- *Configuration Plugin* – Configuration Plugin services are called before the configuration dictionary is given to the configuration targets. The plugin can modify the configuration dictionary.

Figure 27   *Configuration Admin Class Diagram org.osgi.service.cm*

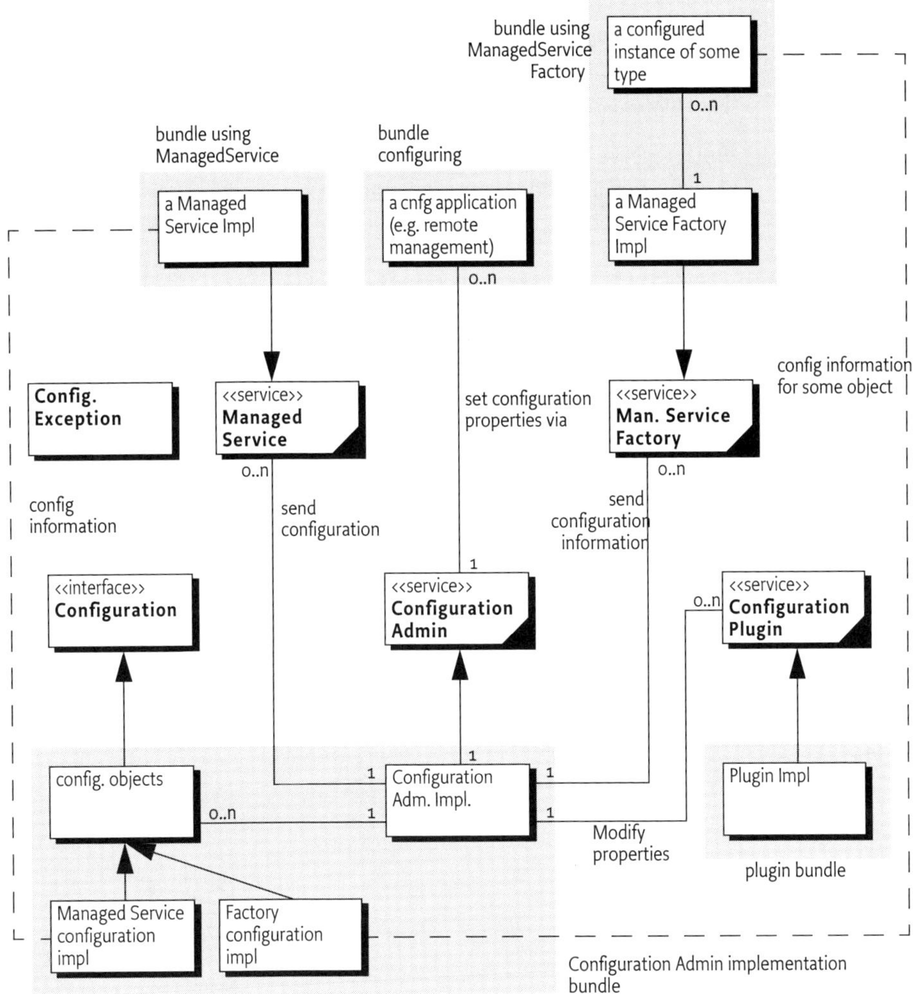

9.2   Configuration Targets

One of the more complicated aspects of this specification is the subtle distinction between the ManagedService and ManagedServiceFactory classes.

Both receive configuration information from the Configuration Admin service and are treated similarly in most respects. Therefore, this specification refers to *configuration targets* when the distinction is irrelevant.

The difference between these types is related to the cardinality of the configuration dictionary. A Managed Service is used when an existing entity needs a configuration dictionary. Thus, a one-to-one relationship always exists between the configuration dictionary and the entity.

A Managed Service Factory is used when part of the configuration is to define *how many instances are required*. A management bundle can create, modify, and delete any number of instances for a Managed Service Factory through the Configuration Admin service. Each instance is configured by a single Configuration object. Therefore, a Managed Service Factory can have multiple associated Configuration objects.

Figure 28   *Differentiation of ManagedService and ManagedServiceFactory Classes*

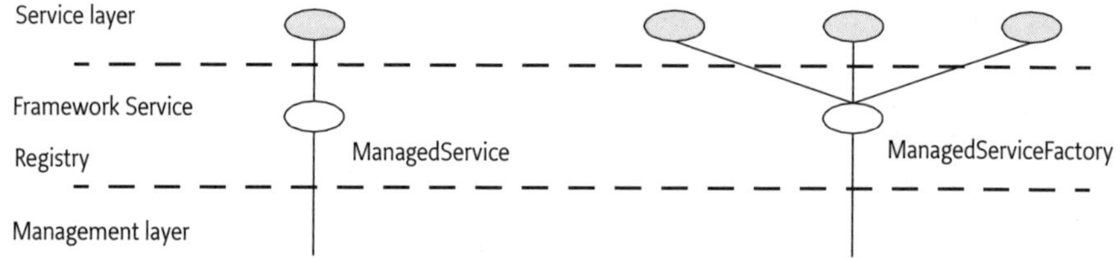

To summarize:

- A *Managed Service* must receive a single configuration dictionary when it is registered or its configuration is modified.
- A *Managed Service Factory* must receive from zero to n configuration dictionaries when it registers, depending on the current configuration. The Managed Service Factory is informed of configuration dictionary changes: modifications, creations, and deletions.

9.3　The Persistent Identity

A crucial concept in the Configuration Admin service specification is the Persistent IDentity (PID). Its purpose is to act as a primary key for objects that need a configuration dictionary. The name of the service property for PID is defined in the Framework in org.osgi.framework.Constants.SERVICE_PID.

A PID is a unique identifier for a service that persists over multiple invocations of the Framework.

When a bundle registers a service with a PID, it should set property service.pid to a unique value. For that service, the same PID should always be used. If the bundle is stopped and later started, the same PID should be used.

PIDs can be useful for all services, but the Configuration Admin service requires their use with Managed Service and Managed Service Factory registrations, because it associates its configuration data with PIDs.

PIDs must be unique for each service. A bundle must not register multiple configuration target services with the same PID. If that should occur, the Configuration Admin service must:

- Send the appropriate configuration data to all services registered under that PID from that bundle only.
- Report an error in the log.
- Ignore duplicate PIDs from other bundles, and report them to the log as well.

9.3.1　PID Readability

PIDs are intended for use by other bundles, not by people, but sometimes the user is confronted with a PID. For example, when installing an alarm system, the user needs to identify the different components to a wiring application. This type of application exposes the PID to end users.

To improve this process, the schemes for PIDs that are defined in this specification should be followed.

Any globally unique string can be used as a PID. The following sections, however, define schemes for common cases. These schemes are not required, but bundle developers are urged to use them to achieve consistency.

9.3.1.1 Local Bundle PIDs

As a convention, descriptions starting with the bundle identity and a dot (.) are reserved for a bundle. As an example, a PID of "65.536" would belong to the bundle with a bundle identity of 65.

9.3.1.2 Software PIDs

Configuration target services that are singletons can use a Java package name they own as the PID (the reverse domain name scheme). As an example, the PID com.acme.watchdog would represent a Watchdog service from the ACME company.

9.3.1.3 Devices

Devices are usually organized on busses or networks. The identity of a device, such as a unique serial number or an address, is a good component of a PID. The format of the serial number should be the same as that printed on the housing or box, to aid in recognition.

Bus	Example	Format	Description
USB	USB-0123-0002-9909873	idVendor (hex 4) idProduct (hex 4) iSerialNumber (decimal)	Universal Serial Bus. Use the standard device descriptor.
IP	IP-172.16.28.21	IP nr (dotted decimal)	Internet Protocol
802	802-00:60:97:00:9A:56	MAC address with : separators	IEEE 802 MAC address (Token Ring, Ethernet, …)
ONE	ONE-06-00000021E461	Family (hex 2) and serial nr including CRC (hex 6)	1-wire bus of Dallas Semiconductor
COM	COM-krups-brewer-12323	serial nr or type name of device	Serial ports

Table 14   *Schemes for Device-Oriented PID Names*

9.4 The Configuration Object

A Configuration object contains the configuration dictionary: a set of properties to configure an aspect of a bundle. A bundle can receive Configuration objects by registering a configuration target service with a PID service property. See *The Persistent Identity* on page 229 for more information about PIDs.

During registration, the Configuration Admin service must detect these targets and hand over their configuration dictionary via a call-back. If this configuration dictionary is subsequently modified, the modified dictionary is handed over to the configuration target again with the same callback.

The Configuration object is primarily a set of properties that can be updated by a management system, user interfaces on the OSGi environment, or other applications. Configuration changes are first made persistent, and then passed to the target service via a call to the updated method in ManagedServiceFactory or ManagedService class.

9.4.1 Location Binding

When a Configuration object is created by either getConfiguration or createFactoryConfiguration, it becomes bound to the location of the calling bundle. This location is obtained with the associated bundle's getLocation method.

Location binding is a security feature that assures that only management bundles can modify configuration data, and other bundles can only modify their own configuration data. A SecurityException is thrown if a non management bundle attempts to modify the configuration information of another bundle.

If a Managed Service is registered with a PID that is already bound to another location, the normal callback to ManagedService.updated must not take place.

Two argument versions of getConfiguration and createFactoryConfiguration take a location String as their second argument. These methods require AdminPermission, and they create Configuration objects bound to the specified location, instead of the location of the calling bundle. These methods are intended for management bundles.

A null location parameter may be used to create Configuration objects that are not bound. In this case, the objects become bound to a specific location the first time that they are used by a bundle.

A management bundle may create a Configuration object before the associated Managed Service is registered. It may use a null location to avoid any dependency on the actual location of the bundle which registers this service. When the Managed Service is registered later, the Configuration object must be bound to the location of the registering bundle, and its configuration dictionary must then be passed to ManagedService.updated.

9.4.2 Configuration Properties

A configuration dictionary contains a set of properties in a Dictionary object. The value types that must be used are the same types as the types supported in the Framework service registry, which are defined as:

```
type        =
        String        | Integer  | Long      | Float
        | Double      | Byte     | Short     | BigInteger
        | BigDecimal  | Character | Boolean
        | vector
        | arrays

primitive =
        long          | int      | short     | char
        | byte        | boolean  | double    | float

arrays      =
    primitive '[]' | type '[]'

vector = Vector of type
```

The name or key of a property must always be a String object, and is not case sensitive during look up, but must preserve the original case. Bundles should not use nested vectors or arrays.

9.4.3 Property Propagation

An implementation of a Managed Service should copy all the properties of the Dictionary object argument in updated(Dictionary), known or unknown, into its service registration properties using ServiceRegistration.setProperties.

This propagation allows the development of applications that leverage the Framework service registry more extensively, so compliance with this mechanism is advised.

A configuration target service may ignore any configuration properties it does not recognize, or it may change the values of the configuration properties before these properties are registered. Configuration properties in the Framework service registry are not strictly related to the configuration information.

Bundles that cooperate with the propagation of configuration properties can participate in horizontal applications. For example, an application that maintains physical location information in the Framework service registry. It could offer support by finding out where a particular device is located in the house or car. This service could use a property dedicated to the physical location and provide functions that leverage this property: for example, a graphic user interface that displays these locations.

9.4.4 Automatic Properties

The Configuration Admin service must automatically add a number of properties to the configuration dictionary. If these properties are set by a configuring bundle, they must always be overridden. Therefore, the receiving bundle can assume that the following properties are defined by the Configuration Admin service and not by the configuring bundle:

- service.pid – Set to be the PID of the associated Configuration object.
- service.factoryPid – Only set for a Managed Service Factory. It is then set to the PID of the associated Managed Service Factory.

Constants for some of these properties can be found in org.osgi.framework.Constants. These system properties are all of type String.

9.5 Managed Service

A Managed Service is used by a bundle that needs one configuration dictionary. Each Managed Service is associated with zero or more Configuration objects in the Configuration Admin service.

A bundle can register any number of ManagedService objects, but each must be identified with its own PID.

A bundle should use a Managed Service when it needs configuration information for:

- *A Singleton* – A single entity in the bundle needs to be configured.
- *Externally Detected Devices* – Each device that is detected causes a registration of an associated ManagedService object. The PID of this object is related to the identity of the device: for example, the address or serial number.

9.5.1 Networks

When a device in the external world needs to be represented in the OSGi Environment, it must be detected in some manner. The control is from the external world, and the Configuration Admin service cannot know the identity and the number of instances without assistance. When a device is detected, it still needs configuration information in order to play a useful role.

For example, a 1-Wire network can automatically detect devices that are attached and removed. For example, when it detects a temperature sensor, it could register a Sensor service with the Framework service registry. This Sensor service needs configuration information specifically for that sensor: which lamps should be turned on, at what temperature it triggers, what timer should be started, in what zone it resides, and so on. One bundle could potentially have hundreds of these sensors and actuators, and each needs its own configuration information.

Each of these Sensor services should be registered as a Managed Service with a PID related to the physical sensor (such as the address) to receive configuration information.

Another examples are services discovered on networks with protocols like Jini, UPnP, and Salutation. They can usually be represented in the Framework service registry. A network printer, for example, could be detected via UPnP. Once in the service registry, these services usually require local configuration information. A Printer service needs to be configured for its local role: location, access list, and so on.

This information needs to be available in the Framework service registry whenever that particular Printer service is registered. Therefore, the Configuration Admin service must remember the configuration information for this Printer service.

This type of service should register with the Framework as a Managed Service in order to receive appropriate configuration information.

9.5.2   Singletons

When an object must be instantiated only once, it is called a *singleton*. A singleton requires a single configuration dictionary. Bundles may implement several different types of singletons if necessary.

A Watchdog service could watch the registry for the status and presence of services in the Framework service registry. Only one instance of a Watchdog service is needed, so only a single configuration dictionary is required, which contains the polling time and the list of services to watch.

9.5.3   Configuring Managed Services

A bundle that needs configuration information should register one or more ManagedService objects with a PID service property. If it has a default set of properties for its configuration, it may include them as service properties of the Managed Service. These properties may be used as a configuration template when a Configuration object is created for the first time. A Managed Service optionally implements the org.osgi.service.metatype.MetaTypeProvider interface to provide information about the property types. See *Meta Typing* on page 250.

When this registration is detected by the Configuration Admin service, the following steps must occur.

- The configuration stored for the registered PID must be retrieved. If there is a Configuration object for this PID, it is sent to the Managed Service with updated(Dictionary).
- If a Managed Service is registered and no configuration information is available, the Configuration Admin service must call updated(Dictionary) with a null parameter.
- If the Configuration Admin service starts up *after* a Managed Service is registered, it must call updated(Dictionary) on this service as soon as possible. For this reason, a Managed Service must always get a callback when it registers *and* the Configuration Admin service is started.

The updated(Dictionary) callback from the Configuration Admin service to the Managed Service must take place on a thread that is different from the one that executed the registration. This requirement allows the Managed Service to finish its initialization in a synchronized method without interference from the Configuration Admin service callback.

Care should be taken not to cause deadlocks by calling the Framework within a synchronized method.

Figure 29   *Managed Service Configuration Action Diagram*

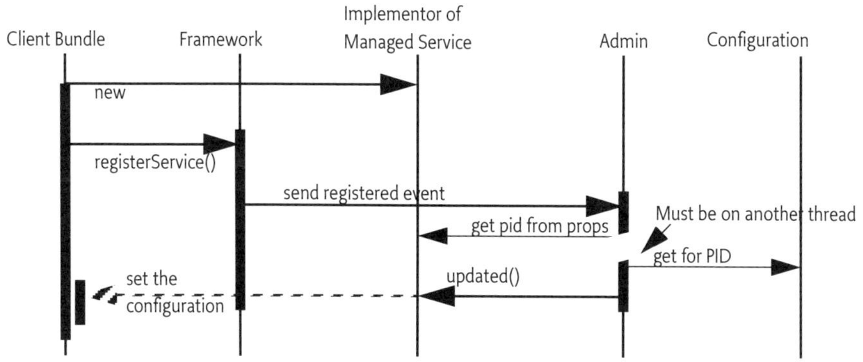

The updated method may throw a ConfigurationException. This object must describe the problem as well as what property caused the exception.

9.5.4 Race Conditions

When a Managed Service is registered, the default properties may be visible in the service registry for a short period before they are replaced by the properties of the actual configuration dictionary. Care should be taken that this visibility does not cause race conditions for other bundles.

In cases where race conditions could be harmful, the Managed Service must be split into two pieces: an object performing the actual service and a Managed Service. First, the Managed Service is registered, and then, after the configuration is received, the actual service object is registered. In such cases, the use of a Managed Service Factory that performs this function more naturally should be considered.

9.5.5 Examples of Managed Service

Figure 30 shows a Managed Service configuration example. Two services are registered under the ManagedService interface, each with a different PID.

Figure 30   *PIDs and External Associations*

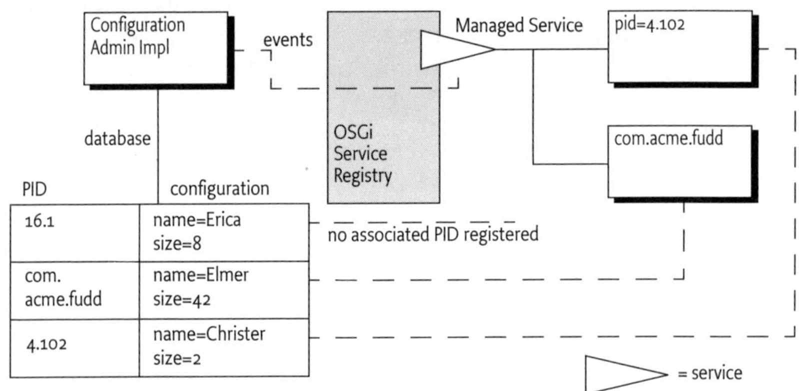

The Configuration Admin service has a database containing a configuration record for each PID. When the Managed Service with service.pid = com.acme.fudd is registered, the Configuration Admin service will retrieve the properties name=Elmer and size=42 from its database. The properties are stored in a Dictionary object and then given to the Managed Service with the updated(Dictionary) method.

9.5.5.1 Configuring A Console Bundle

In this example, a bundle can run a single debugging console over a Telnet connection. It is a singleton, so it uses a ManagedService object to get its configuration information: the port and the network name on which it should register.

```
class SampleManagedService implements ManagedService {
  Dictionary            properties;
  ServiceRegistration   registration;
  Console               console;

  public synchronized void start(
    BundleContext context ) throws Exception {
    properties = new Hashtable();
    properties.put( Constants.SERVICE_PID,
      "com.acme.console" );
    properties.put( "port",  new Integer(2011) );

    registration = context.registerService(
      ManagedService.class.getName(),
      this,
      properties
    );
  }

  public synchronized void updated( Dictionary np ) {
    if ( np != null ) {
      properties = np;
      properties.put(
        Constants.SERVICE_PID, "com.acme.console" );
    }

    if (console == null)
      console = new Console();

    int port = ((Integer)properties.get(
"port")).intValue();

    String network = (String) properties.get(
"network");
    console.setPort(port, network);
    registration.setProperties(properties);
  }
  ... further methods
}
```

9.5.6   Deletion

When a Configuration object for a Managed Service is deleted, the
Configuration Admin service must call updated(Dictionary) with a
null argument on a thread that is different from that on which the
Configuration.delete was executed.

9.6   Managed Service Factory

A Managed Service Factory is used when configuration is needed for a service that can be instantiated multiple times. When a Managed Service Factory is registered with the Framework, the Configuration Admin service consults its database and call updated(String, Dictionary) for each associated Configuration object. It passes the identifier of the instance, which can be used as a PID, as well as a Dictionary object with the configuration properties.

A Managed Service Factory is useful when the bundle can provide functionality a number of times, each time with different configuration dictionaries. In this situation, the Managed Service Factory acts like a *class* and the Configuration Admin service can use this Managed Service Factory to *instantiate instances* for that *class*.

In the next section, the word *factory* refers to this concept of creating *instances* of a function defined by a bundle that registers a Managed Service Factory.

9.6.1   When to Use a Managed Service Factory

A Managed Service Factory should be used when the bundle does not have an internal or external entity associated with the configuration information, but can potentially be instantiated multiple times.

9.6.1.1   Example Email Fetcher

An email fetcher program shows the number of emails that a user has on a display – a function likely to be required for different users. This function could be viewed as a *class* that needs to be *instantiated* for each user. Each instance requires different parameters: password, host, protocol, user id, and so on.

An implementation of the Email Fetcher service should register a ManagedServiceFactory object. In this way, the Configuration Admin service can define the configuration information for each user separately. The Email Fetcher service will only receive a configuration dictionary for each required instance (user).

9.6.1.2   Example Temperature Conversion Service

A bundle has the code to implement a conversion service that receives a temperature, and, depending on settings, can turn an actuator on and off. This service would need to be instantiated many times depending on where it is needed. Each instance would require its own configuration information for:

- Upper value
- Lower value

- Switch Identification
- ...

Such a conversion service should register a service object under a ManagedServiceFactory interface. A configuration program can then use this Managed Service Factory to create instances as needed. For example, this program could use a Graphic User Interface (GUI) to create such a component and configure it.

9.6.1.3 Serial Ports

Serial ports cannot always be used by the OSGi Device Access specification implementations. First, some environments have no means to identify available serial ports. Second, a device on a serial port cannot always provide information about its type.

Therefore, each serial port requires a description of the device that is connected. The bundle managing the serial ports would need to instantiate a number of serial ports under control of the Configuration Admin service, with the appropriate DEVICE_CATEGORY property to allow it to participate in the Device Access implementation.

If the bundle cannot detect the available serial ports automatically, it should register a Managed Service Factory. The Configuration Admin service can then, with the help of a configuration program, define configuration information for each available serial port.

9.6.2 Registration

The configuration dictionary for a Managed Service Factory is identified by a PID, similar to the Managed Service configuration dictionary. The Managed Service Factory, however, also has a *factory* PID: the PID of the associated Managed Service Factory. It is used to group all Managed Service Factory configuration dictionaries together.

When a Configuration object for a factory is created (Configuration-Admin.createFactoryConfiguration), a new unique PID is created for this object by the Configuration Admin service. The scheme used for this PID is defined by the Configuration Admin service, and is unrelated to the factory PID.

When the Configuration Admin service detects the registration of a Managed Service Factory, it must find all configuration dictionaries for this factory and must then sequentially call ManagedService-Factory. updated(String, Dictionary) for each configuration dictionary. The first argument is the PID of the Configuration object (the one created by the Configuration Admin service) and the second argument contains the configuration properties.

The receiver should then create instances of the associated factory class. The bundle may register new services with the Framework, using the PID given in the Configuration object, but it is not required. The configuration dictionary may be used internally only.

The Configuration Admin service must guarantee that the Configuration objects are not deleted before their properties are given to the Managed Service Factory, and must assure that no race conditions exist between initialization and updates.

Figure 31   *Managed Service Factory Action Diagram*

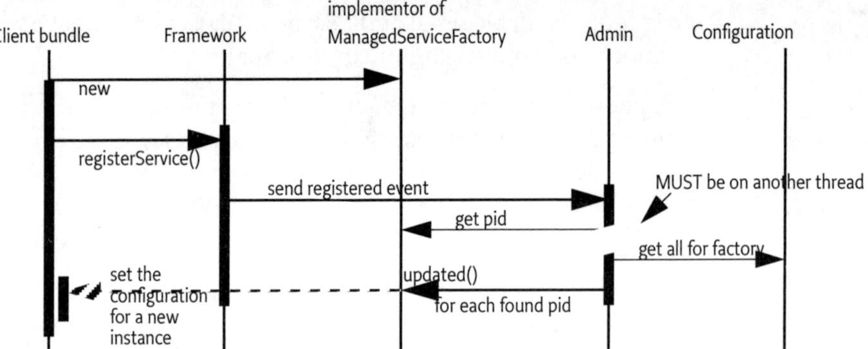

A Managed Service Factory has only one update method: updated(String,Dictionary). This method can be called any number of times as Configuration objects are created or updated.

The Managed Service Factory must detect whether a PID is being used for the first time, in which case it should create a new *instance*, or a subsequent time, in which case it should update an existing instance.

The Configuration Admin service must call updated(String,Dictionary) on a thread that is different from the one that executed the registration. This requirement allows an implementation of a Managed Service Factory to use a synchronized method to assure that the callbacks do not interfere with the Managed Service Factory registration.

The updated(String,Dictionary) method may throw a ConfigurationException object. This object describes what property caused the problem and what the problem was. These exceptions should be logged by a Configuration Admin service.

9.6.3   Deletion

If a configuring bundle deletes an instance of a Managed Service Factory, the deleted(String) method is called. The argument is the PID for this instance. The implementation of the Managed Service Factory must remove all information, and stop any behavior, associated with that PID. If a service was registered for this PID, it should be unregistered.

9.6.4   Example of Managed Service Factory

Figure 32 highlights the differences between a Managed Service and a Managed Service Factory. It shows how a Managed Service Factory implementation receives the configuration information that was created before it was registered.

- A bundle implements an EMail Fetcher service. It registers a ManagedServiceFactory object with PID=com.acme.email.
- The Configuration Admin service notices the registration and consults its database. It finds three Configuration objects for which the factory PID is equal to com.acme.email. It must call updated(String,Dictionary) for each of these Configuration objects, on the newly registered ManagedServiceFactory object.
- For each configuration dictionary received, the factory should create a new instance of a EMailFetcher object, one for erica (PID=16.1), one for anna (PID=16.3) and one for elmer (PID=16.2).
- The EMailFetcher objects are registered under the Topic interface so their results can be viewed by an online display.
  If the MailFetcher object is registered, it may safely use the PID of the Configuration object because the Configuration Admin service must guarantee its suitability for this purpose.

Figure 32   *Managed Service Factory Example*

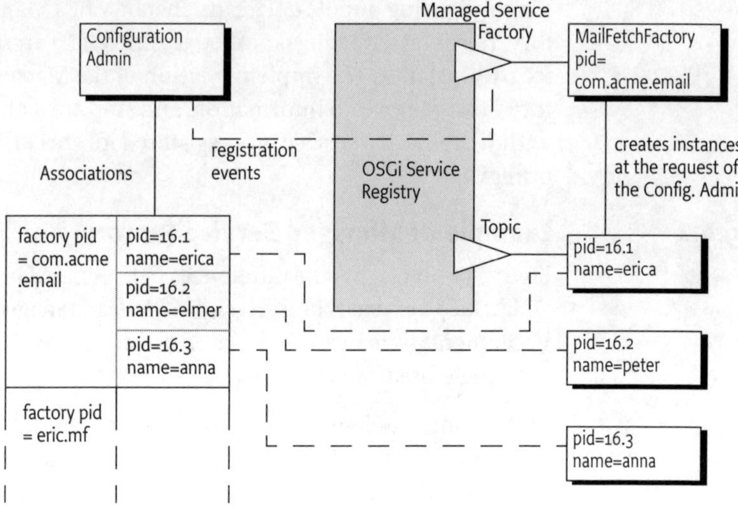

9.6.5   Multiple Consoles Example

This example allows multiple consoles, each of which has its own port and interface, to run simultaneously. This approach is very similar to the example for the Managed Service, but highlights the difference by allowing multiple consoles to be created.

```
class ExampleFactory implements ManagedServiceFactory
{
   Hashtable consoles = new Hashtable();

   public void start( BundleContext context )
      throws Exception {
      Hashtable local = new Hashtable();
      local.put( Constants.SERVICE_PID,
"com.acme.console" );
      context.registerService(
         ManagedServiceFactory.class.getName(),
         this,
         local
      );
   }

   public void updated( String pid, Dictionary config
){
       Console console = (Console) consoles.get(pid);
       if (console == null) {
```

```
          console = new Console(context);
          consoles.put(pid, console);
        }

        int port = getInt(config, "port", 2011);
        String network = getString(
          config,
          "network",
          null /*all*/
        );
        console.setPort(port, network);
      }

    public void deleted(String pid) {
        console = (Console) consoles.get(pid);
        if (console != null) {
          consoles.remove(pid);
          console.close();
        }
      }
    }
  }
```

9.7   Configuration Admin Service

The ConfigurationAdmin interface provides methods to maintain
the configuration data in an OSGi environment. This configuration
information is defined by a number of Configuration objects associated with specific configuration targets. Configuration objects can
be created, listed, modified, and deleted through this interface.
Either a remote management system or the bundles configuring
their own configuration information may perform these operations.

The ConfigurationAdmin interface has methods for creating and
accessing the Configuration objects for a Managed Service service, as
well as methods for managing new Configuration objects for a Managed Service Factory.

9.7.1   Creating a Managed Service Configuration Object

A bundle can create a new Managed Service Configuration object
with ConfigurationAdmin.getConfiguration. No create method is
offered because doing so could introduce race conditions between
different bundles creating the same Configuration object. The get-
Configuration method must atomically create and persistently store
an object if it does not yet exist.

Two variants of this method are:

- getConfiguration(String) – This method is used by a bundle with a given location to configure its *own* ManagedService objects. The argument specifies the PID of the targeted service.
- getConfiguration(String, String) – This method is used by a management bundle to configure *another* bundle. Therefore, this management bundle needs AdminPermission. The first argument is the location identifier and the second is the PID of the targeted ManagedService object.

All Configuration objects have a method, getFactoryPid(), which in this case must return null because the Configuration object is associated with a Managed Service.

9.7.2 Creating a Managed Service Factory Configuration Object

The ConfigurationAdmin class provides two methods to create a new instance of a Managed Service Factory:

- createFactoryConfiguration(String) – This method is used by a bundle with a given location to configure its own ManagedServiceFactory objects. The argument specifies the PID of the targeted ManagedServiceFactory object. This *factory PID* can be obtained from the returned Configuration object with the get-FactoryPid() method.
- createFactoryConfiguration(String, String)– This method is used by a management bundle to configure another bundle's ManagedServiceFactory object. This management bundle needs AdminPermission. The first argument is the location identifier and the second is the PID of the targeted ManagedServiceFactory object. This *factory PID* can be obtained from the returned Configuration object with getFactoryPid method.

9.7.3 Accessing Existing Configurations

The existing set of Configuration objects can be listed with the listConfigurations(String). The argument is a String object with a filter expression. This filter expression has the same syntax as the Framework Filter. For example:

```
(&(size=42)(service.bundleLocation=*osgi*))
```

The filter function must use the properties of the Configuration objects and only return the ones that matches the filter expression.

A single Configuration object is identified with a PID and can be obtained with getConfiguration(String).

If the caller has AdminPermission, then all Configuration objects are eligible for search. In other cases, only Configuration objects bound to the calling bundle's location must be returned.

null is returned in both cases when an appropriate Configuration object cannot be found.

9.7.3.1 Updating a Configuration

The process of updating a Configuration object is the same for Managed Services and Managed Service Factory. First, listConfigurations(String) or getConfiguration(String) should be used to get a Configuration object. The properties can be obtained with Configuration.getProperties. When no update has occured since this object was created, getProperties returns null.

New properties can be set by calling Configuration.update. The Configuration Admin service first stores the configuration information and then calls the configuration target updated method: either the ManagedService.updated or ManagedServiceFactory.updated method. If this target service is not registered, the fresh configuration information must be set when the configuration target service registers.

The update method calls in Configuration objects are not executed synchronously with the related target service updated method. This method is called on a different thread, some period of time after the configuration information has been updated in the Configuration object. The Configuration Admin service, however, must have updated the persistent storage before the update method returns.

9.7.4 Deletion

A Configuration object that is no longer needed can be deleted with Configuration.delete, which removes the Configuration object from the database. The database must be updated before the target service updated method is called.

If the target service is a Managed Service Factory, the factory is informed of the deleted Configuration object by a call to ManagedServiceFactory.deleted. It should then remove the associated *instance*. The ManagedServiceFactory.deleted call is done on a separate thread some period of time after the Configuration object has been deleted, and hence is asynchronous with respect to Configuration.delete.

When a Configuration object of a Managed Service is deleted, ManagedService.updated is called with null for the properties argument. This method may be used for cleanup, to revert to default values, or to stop servicing.

9.7.5 **Updating a Bundle's Own Configuration**

The Configuration Admin service specification does not distinguish between updates via a management bundle and a bundle updating its own configuration information (as defined by its location). Even if a bundle updates its own configuration information, the Configuration Admin service must callback the associated target service updated method.

As a rule, to update its own configuration, a bundle's user interface should *only* update the configuration information and never its internal structures directly. This rule has the advantage that the events, from the bundle implementation's perspective, appear similar for internal updates, remote management updates, and initialization.

9.8 Configuration Plugin

The Configuration Admin service allows third-party applications to participate in the configuration process; bundles that register a service object under a ConfigurationPlugin interface can process the configuration dictionary before it reaches the configuration target service.

Plugins allow sufficiently privileged bundles to intercept configuration dictionaries before they are passed to the intended Managed Service or Managed Service Factory. The ConfigurationPlugin interface has only one method: modifyConfiguration(ServiceReference, Dictionary) – This method inspects or modifies the configuration data.

All plugins in the service registry are traversed and called before the properties are passed to the configuration target service. Each Configuration Plugin object gets a chance to inspect the existing data, look at the target object which can be a ManagedService object or a ManagedServiceFactory object, and modify the properties of the configuration dictionary.

Obviously, ConfigurationPlugin objects should not modify properties that belong to the configuration properties of the target service unless the implications are understood.This functionality is mainly intended to provide functions that leverage the Framework service registry.

For example, a Configuration Plugin service may add a physical location property to a service. This property can be leveraged by applications that want to know where a service is physically located. This scenario could be carried out without any further support of the service itself, except for the general requirement that the service should propagate the properties it receives from the Configuration Admin service to the service registry.

Figure 33   *Order of Configuration Plugin Services*

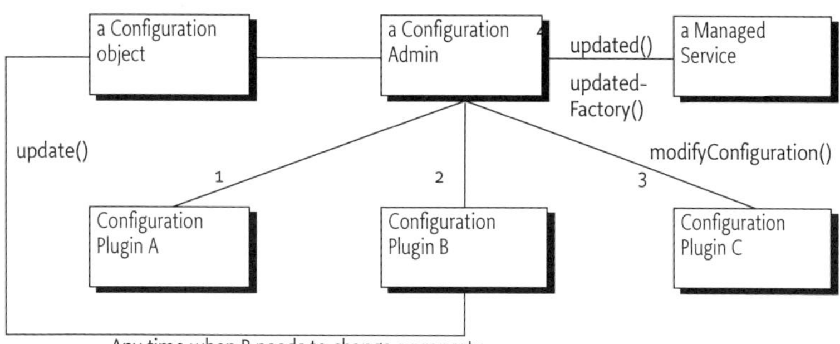

9.8.1   Limiting The Targets

A ConfigurationPlugin object may optionally specify a cm.target registration property. The value is the PID of the configuration target whose configuration updates the ConfigurationPlugin objects wants to intercept.

The ConfigurationPlugin object must then only be called with updates for the configuration target service with the specified PID. Omitting the cm.target registration property means that it is called for *all* configuration updates.

9.8.2   Example of Property Expansion

Consider a Managed Service that has a configuration property service.to with the value (objectclass=com.acme.Alarm). When the Configuration Admin service sets this property on the target service, a ConfigurationPlugin object may replace the (object-class=com.acme.Alarm) filter with an array of existing alarm systems' PIDs:

```
ID "service.to=[32434,232,12421,1212]"
```

Now a new Alarm Service with service.pid=343 is registered, requiring that the to list of the target service be updated. The bundle which registered the Configuration Plugin service, therefore, wants to set the to registration property on the target service. It does *not* do this by calling ManagedService.updated directly for several reasons:

- In a securely configured system it should not have the permission to make this call or even obtain the target service.
- It could get into race conditions with the Configuration Admin service if it did. Both services would compete for access simultaneously.

Instead, it must get the Configuration object from the Configuration Admin service and call the update method on it.

The Configuration Admin service must schedule a new update cycle on another thread, and sometime in the future must call ConfigurationPlugin.modifyProperties. The ConfigurationPlugin object could then set the service.to property to [32434,232,12421,1212, 343]. After that, the Configuration Admin service must call updated on the target service with the new service.to list.

9.8.3　Configuration Data Modifications

Modifications to the configuration dictionary are still under control of the Configuration Admin service, which must determine whether to accept these changes, hide critical variables, or deny changes for other reasons.

The ConfigurationPlugin interface must also allow plugins to detect configuration updates to the service via the callback. This ability allows them to synchronize the configuration updates with transient information.

9.8.4　Forcing a Callback

If a bundle needs to force a Configuration Plugin service to be called again, it must fetch the appropriate Configuration object from the Configuration Admin service and call the update() method (the no parameter version) on this object. This call forces an update with the current configuration dictionary, so that all applicable plugins get called again.

9.8.5　Calling Order

The order in which the ConfigurationPlugin objects are called must depend on the service.cmRanking configuration property of the ConfigurationPlugin object. Table 15 shows the usage of the service.cmRanking property for order of calling the Configuration Plugin services.

service.cmRanking value	Description
< 0	The Configuration Plugin service should not modify properties and must be called before any modifications are made.
> 0 && <= 1000	Modifies the configuration dictions. The calling order should be based on the value of the service.cmRanking property.
> 1000	Should not modify data and shall be called after all modifications are made.

Table 15   service.cmRanking *Usage For Ordering*

9.9   Remote Management

This specification does not attempt to define a remote management interface for the Framework. The purpose of this specification is to define a minimal interface for bundles which is complete enough to test.

The Configuration Admin service is a primary aspect of remote management, however, and this specification must be compatible with common remote management standards. This section discusses some of the issues of using this specification with [26] *DTMF Common Information Model* (CIM) and [27] *Simple Network Management Protocol* (SNMP), the most likely candidates for remote management today.

These discussions are not complete, comprehensive, or normative. They are intended to point the bundle developer in relevant directions. Further specifications are needed to make a more concrete mapping.

9.9.1   Common Information Model

Common Information Model (CIM) defines the managed objects in [29] *Interface Definition Language* (IDL) language which was developed for the Common Object Request Broker Architecture (CORBA).

The data types have a syntax, and the data values have a syntax. Additionally these syntaxes can be mapped to XML. Unfortunately, this XML mapping is very different from the very applicable [28] *XSchema* XML data type definition language. The Framework service registry property types are a proper subset of the CIM data types.

In this specification, a Managed Service Factory would map to a CIM class definition. The primitives create, delete, and set are supported in this Specification via the ManagedServiceFactory interface. The possible data types in CIM are richer than the Framework supports, and should thus be limited to cases when CIM classes for bundles are defined.

An important conceptual difference between this specification and CIM is the naming of properties. CIM properties are scoped inside a class. In this Specification, properties are primarily scoped inside the Managed Service Factory, but are then placed in the registry, where they have global scope. This mechanism is like [30] *Lightweight Directory Access Protocol*, in which the semantics of the properties are defined globally and a class is a collection of globally defined properties.

The specification does not address the non-Configuration Admin service primitives, such as notifications and method calls.

9.9.2 Simple Network Management Protocol

The Simple Network Management Protocol (SNMP) defines the data model in ASN.1. SNMP is a rich data typing language that supports many types that are difficult to map to the data types supported in this specification. A large overlap exists, however, and it should be possible to design a data type that is applicable in this context.

In this case, the PID of a Managed Service should map to the SNMP Object IDentifier (OID). Managed Service Factories are mapped to tables in SNMP, although this mapping creates an obvious restriction in data types because tables can only contain scalar values. Therefore, the property values of the Configuration object would have to be limited to scalar values.

Similar scoping issues as seen in CIM arise for SNMP because properties have a global scope in the service registry.

SNMP does not support the concept of method calls or function calls. All information is conveyed as the setting of values. The SNMP paradigm maps closely to this specification.

This specification does not address the non-Configuration Admin primitives, such as traps.

9.10 Meta Typing

This section discusses how the Metatype specification is used in the context of a Configuration Admin service.

When a Managed Service or Managed Service Factory is registered, the service object may also implement the MetaTypeProvider interface.

If the Managed Service or Managed Service Factory service object implements the MetaTypeProvider interface, a management bundle may assume that the associated ObjectClassDefinition object can be used to configure the service.

The ObjectClassDefinition and AttributeDefinition objects contain sufficient information to automatically build simple user interfaces. They can also be used to augment dedicated interfaces with accurate validations.

This specification does not address how the metatype is made available to a management system, due to the many open issues regarding remote management.

9.11 Security

9.11.1 Permissions

Configuration Admin service security is implemented using ServicePermission and AdminPermission. The following table summarizes the permissions needed by the Configuration Admin bundle itself, as well as those needed by the bundles with which it interacts.

Bundle Registering	ServicePermisson Action	AdminPermission
ConfigurationAdmin	REGISTER ConfigurationAdmin GET ManagedService GET ManagedServiceFactory GET ConfigurationPlugin	Yes
ManagedService	REGISTER ManagedService GET ConfigurationAdmin	No
ManagedServiceFactory	REGISTER ManagedServiceFactory GET ConfigurationAdmin	No
ConfigurationPlugin	REGISTER ConfigurationPlugin GET ConfigurationAdmin	No

Table 16   *Permission Overview Configuration Admin*

The Configuration Admin service must have ServicePermission[REGISTER], for the ConfigurationAdmin interface. It will also be the only bundle that needs the ServicePermission[GET], for ManagedService, ManagedServiceFactory and ConfigurationPlugin interfaces. No other bundle should be allowed to have GET permission for these interfaces. The Configuration Admin bundle must also hold AdminPermission.

Bundles that can be configured must have the ServicePermission[REGISTER], for ManagedService or ManagedServiceFactory interfaces.

Bundles registering ConfigurationPlugin objects must have the ServicePermission with action REGISTER for ConfigurationPlugin interfaces. The Configuration Admin service must trust all services registered with the ConfigurationPlugin interface. Only the Configuration Admin service should have ServicePermission[GET] on a ConfigurationPlugin interface.

If a Managed Service or Managed Service Factory is implemented in an object that is also registered under another interface, it is possible, although inappropriate, for a bundle other than the Configuration Admin service implementation to call updated. Security-aware bundles can avoid this problem by having their updated methods check that the caller holds the AdminPermission.

Bundles that want to change their own configuration need ServicePermission[GET] for ConfigurationAdmin interface. A bundle with AdminPermission is allowed to access and modify any Configuration object.

Pre-configuration of bundles requires AdminPermission because the methods that specify a location require this permission.

9.11.2 Forging PIDs

A risk exists of an unauthorized bundle forging a PID in order to obtain and possibly modify the configuration information of another bundle. To mitigate this risk, Configuration objects are generally *bound* to a specific bundle location, and are not passed to any Managed Service or Managed Service Factory registered by a different bundle.

Bundles with the required AdminPermission can create Configuration objects that are not bound: that is, they have their location set to null. This ability can be useful for pre-configuring bundles before they are installed, without having to know their actual locations.

In this scenario, the Configuration object must become bound to the first bundle which registers a Managed Service (or Managed Service Factory) with the right PID.

A bundle could still possibly obtain another bundle's configuration, by registering a Managed Service with the right PID before the victim bundle does so. This situation can be regarded as a denial-of-service attack, because the victim bundle would never receive its configuration information. Such an attack can be avoided by always binding Configuration objects to the right locations. It can also be detected by the Configuration Admin service, because later, when the victim bundle registers the correct PID, two equal PIDs are registered, and this violation should be logged.

9.11.3 Configuration and Permission Administration

Configuration information has a direct influence on the permissions needed by a bundle. For example, when the Configuration Admin Bundle orders a bundle to use port 2011 for a console, that bundle also needs permission for listening to incoming connections on that port.

Both a simple and a complex solution are available for this situation.

The simple solution is when the bundle has a set of permissions that do not define specific values but allow a range of values. For example, a bundle could listen to ports above 1024 freely. All these ports could then be used for configuration.

The other solution is more complicated. In an environment where there is very strong security, the bundle would only be allowed a specific port. This situation requires an atomic update of both the configuration data and the permissions. If this update was not atomic, a potential security hole would exist during the period of time that the set of permissions did not match the configuration.

The following scenario can be used to update a configuration and the security permissions.

- Stop the bundle.
- Update the appropriate Configuration object via the Configuration Admin service.
- Update the permissions in the Framework.
- Start the bundle.

This scenario would achieve atomicity from the point of view of the bundle.

9.12   Configurable Service

Both the Configuration Admin service and the org.osgi.framework.Configurable interface address configuration management issues. It is the intention of this specification to replace this Framework interface for configuration management.

The Framework Configurable mechanism works as follows. A registered service object implements the Configurable interface to allow a management bundle to configure that service. The Configurable interface has only one method: getConfigurationObject(). This method returns a Java Bean. Beans can be examined and modified with the java.reflect or java.bean packages.

This scheme has the following disadvantages:

- *No factory* – Only registered services can be modified, unlike the Managed Service Factory that creates any number of services.
- *Atomicity* – The beans or reflection API can only modify one property at a time, and there is no way to tell the bean that no more modifications to the properties will follow. This limitation complicates updates of configurations that have dependencies between properties.
  This specification passes a Dictionary object that contains all the configuration properties at once.
- *Profile* – The Java beans API is linked to many packages that are not likely to be present in OSGi environments. The reflection API may be present, but is not simple to use.
  This specification has no required libraries.
- *User Interface support* – UI support in beans is very rudimentary when no AWT is present.
  The associated Metatyping specification does not require any external libraries, and has extensive support for UIs including localization.

9.13   org.osgi.service.cm

The OSGi Configuration Admin service Package. Specification Version 1.0.

Bundles wishing to use this package must list the package in the Import-Package header of the bundle's manifest. For example:

```
Import-Package: org.osgi.service.cm; specification-
version=1.0
```

Class Summary

Interfaces

Configuration	The configuration information for a Managed Service or Managed Service Factory object.
ConfigurationAdmin	Service for administering configuration data.
ConfigurationPlugin	A service interface for processing configuration dictionary before the update.
ManagedService	A service that can receive configuration data from a Configuration Admin service.
ManagedServiceFactory	Manage multiple service instances.

Exceptions

ConfigurationException	An Exception class to inform the Configuration Admin service of problems with configuration data.

9.13.1   Configuration

public interface Configuration

The configuration information for a Managed Service or Managed-Service Factory object. The Configuration Admin service uses this interface to represent the configuration information for a Managed-Service or for a service instance of a Managed Service Factory.

A Configuration object contains a set of configuration dictionary and allows the properties to be updated via this object. Bundles wishing to receive configuration dictionaries do not need to use this class - they register a Managed Service or Managed Service Factory. Only administrative bundles, and bundles wishing to update their own configurations need to use this class.

The properties handled in this configuration have case insensitive String objects as keys. However, case is preserved from the last set key/value. The value of the property may be of the following types:

```
type        = String | Integer | Long
  | Float | Double | Byte
  | Short | BigInteger
  | BigDecimal | Character
  | Boolean | vector | arrays
primitive   = long | int | short
  | char | byte | double | float
arrays      = primitive '[]' | type '[]'
```

```
vector        = Vector of type
```

A configuration can be *bound* to a bundle location (Bundle.get-Location()). The purpose of binding a Configuration object to a location is to make it impossible for another bundle to forge a PID that would match this configuration. When a configuration is bound to a specific location, and a bundle with a different location registers a corresponding Managed Service object or Managed Service Factory object, then the configuration is not passed to the updated method of that object.

If a configuration's location is null, it is not yet bound to a location. It will become bound to the location of the first bundle that registers a Managed Service or Managed Service Factory object with the corresponding PID.

The same Configuration object is used for configuring both a Managed Service Factory and a Managed Service. When it is important to differentiate between these two the term "factory configuration" is used.

9.13.1.1   Methods

public void **delete**()
    throws IOException, IllegalStateException

Delete this Configuration object. Removes this configuration object from the persistent store. Notify on a different thread the corresponding Managed Service or Managed Service Factory. A Managed-Service object is notified by a call to its updated method with a null properties argument. A Managed Service Factory object is notified by a call to its deleted method.

Throws: IOException - If delete fails

IllegalStateException - if this configuration has been deleted

public java.lang.String **getBundleLocation**()
    throws SecurityException, IllegalStateException

Get the bundle location. Returns the bundle location to which this configuration is bound, or null if it is not yet bound to a bundle location.

This call requires Admin Permission.

Returns: location to which this configuration is bound, or null.

Throws: SecurityException - if the caller does not have Admin Permission.

IllegalStateException - if this Configuration object has been deleted.

public java.lang.String **getFactoryPid**()
   throws IllegalStateException

For a factory configuration return the PID of the corresponding Managed Service Factory, else return null.

Returns: factory PID or null

Throws: IllegalStateException - if this configuration has been deleted

public java.lang.String **getPid**()
   throws IllegalStateException

Get the PID for this Configuration object.

Returns: the PID for this Configuration object.

Throws: IllegalStateException - if this configuration has been deleted

public java.util.Dictionary **getProperties**()
   throws IllegalStateException

Return the properties of this Configuration object. The Dictionary object returned is a private copy for the caller and may be changed without influencing the stored configuration. The keys in the returned dictionary are case insensitive and are always of type String.

If called just after the configuration is created and before update has been called, this method returns null.

Returns: A private copy of the properties for the caller or null.

Throws: IllegalStateException - if this configuration has been deleted

public void **setBundleLocation**(java.lang.String bundleLocation)
   throws SecurityException, IllegalStateException

Bind this Configuration object to the specified bundle location. If the bundleLocation parameter is null then the Configuration object will not be bound to a location. It will be set to the bundle's location before the first time a Managed Service/Managed Service Factory receives this Configuration object via the updated method and before any plugins are called.

This method requires AdminPermission.

Parameters: bundleLocation - a bundle location or null

Throws: SecurityException - if the caller does not have AdminPermission

              IllegalStateException - if this configuration has been deleted

public void **update**()
    throws IOException, IllegalStateException

Update the Configuration object with the current properties. Initate the updated callback to the Managed Service or Managed Service Factory with the current properties on a different thread.

This is the only way for a bundle that uses a Configuration Plugin service to initate a callback. For example, when that bundle detects a change that requires an update of the Managed Service or Managed Service Factory via its ConfigurationPlugin object.

Throws: IOException - if update cannot access the properties in persistent storage

IllegalStateException - if this configuration has been deleted

See Also: ConfigurationPlugin

public void **update**(java.util.Dictionary properties)
    throws IOException, IllegalArgumentException,
    IllegalStateException

Update the properties of this Configuration object. Stores the properties in persistent storage after adding or overwriting the following properties:

- "service.pid" : is set to be the PID of this configuration.
- "service.factoryPid" : if this is a factory configuration it is set to the factory PID else it is not set.
- "service.bundleLocation" is set to the location to which this Configuration object is bound. If the location is null, this property is not set.

These system properties are all of type String.

If the corresponding Managed Service/Managed Service Factory is registered, its updated method will be called on another thread. Else, this callback is delayed until aforementioned registration occurs.

Parameters: properties - the new set of properties for this configuration

Throws: IOException - if update cannot be made persistent

IllegalArgumentException - if the Dictionary object contains invalid configuration types

IllegalStateException - if this configuration has been deleted

9.13.2   ConfigurationAdmin

public interface ConfigurationAdmin

Service for administering configuration data.

The main purpose of this interface is to store bundle configuration data persistently. This information is represented in Configuration objects. The actual configuration data is a Dictionary of properties inside a Configuration object.

There are two principally different ways to manage configurations. First there is the concept of a Managed Service, where configuration data is uniquely associated with an object registered with the service registry.

Next, there is the concept of a factory where the Configuration Admin service will maintain o or more Configuration objects for a Managed Service Factory that is registered with the Framework.

The first concept is intended for configuration data about "things/ services" whose existence is defined externally, e.g. a specific printer. Factories are intended for "things/services" that can be created any number of times, e.g. a configuration for a DHCP server for different networks.

Bundles that require configuration should register a Managed Service or a Managed Service Factory in the service registry. A registration property named service.pid (persistent identifier or PID) must be used to identify this Managed Service or Managed Service Factory to the Configuration Admin service.

When the ConfigurationAdmin detects the registration of a Managed Service, it checks its persistent storage for a configuration object whose PID matches the PID registration property (service.pid) of the Managed Service. If found, it calls updated(Dictionary) method with the new properties. The implementation of a Configuration Admin service must run these call-backs one a thread that differs from the initiating thread to allow proper synchronization.

When the Configuration Admin service detects a Managed Service Factory registration, it checks its storage for configuration objects whose factoryPid matches the PID of the Managed Service Factory. For each such Configuration objects, it calls the ManagedService-Factory.updated method on a different thread with the new properties. The calls to the updated method of a Managed Service Factory must be executed sequential and not overlap in time.

In general, bundles having permission to use the Configuration Admin service can only access and modify their own configuration information. Accessing or modifying the configuration of another bundle requires AdminPermission.

Configuration objects can be *bound* to a specified bundle location. In this case, if a matching Managed Service or Managed Service Factory is registered by a bundle with a different location, then the Configuration Admin service must not do the normal callback, and it should log an error. In the case where a Configuration object is not bound, its location field is null, the Configuration Admin service will bind it to the location of the bundle that registers the first Managed Service or Managed Service Factory that has a corresponding PID property.

The method descriptions of this class refer to a concept of "the calling bundle". This is a loose way of referring to the bundle which obtained the Configuration Admin service from the service registry. Implementations of ConfigurationAdmin must use a ServiceFactory to support this concept.

9.13.2.1 Methods

public Configuration **createFactoryConfiguration**(java.lang.String factoryPid)
    throws IOException, SecurityException

Create a new factory Configuration object with a new PID. The properties of the new Configuration object are null until the first time that its update(Dictionary) method is called.

It is not required that the factoryPid maps to a registered Managed Service Factory.

The Configuration object is bound to the location of the calling bundle.

Parameters: factoryPid - PID of factory (not null).

Returns: a new Configuration object.

Throws: IOException - if access to persistent storage fails.

SecurityException - if caller does not have AdminPermission and factoryPid is bound to another bundle.

public Configuration **createFactoryConfiguration**(java.lang.String factoryPid, java.lang.String location)
    throws IOException, SecurityException

Create a new factory Configuration object with a new PID. The properties of the new Configuration object are null until the first time that its update(Dictionary) method is called.

It is not required that the factory Pid maps to a registered Managed Service Factory.

The Configuration is bound to the location specified. If this location is null it will be bound to the location of the first bundle that registers a Managed Service Factory with a corresponding PID.

This method requires Admin Permission.

Parameters: factoryPid - PID of factory (not null).

location - a bundle location string, or null.

Returns: a new Configuration object.

Throws: IOException - if access to persistent storage fails.

SecurityException - if caller does not have Admin Permission.

public Configuration **getConfiguration**(java.lang.String pid)
    throws IOException, SecurityException

Get an existing or new Configuration object from the persistent store. If the Configuration object for this PID does not exist, create a new Configuration object for that PID, where properties are null. Bind its location to the calling bundle's location.

Else, if the location of the existing Configuration object is null, set it to the calling bundle's location.

If the location of the Configuration object does not match the calling bundle, throw a Security Exception.

Parameters: pid - persistent identifier.

Returns: an existing or new Configuration matching the PID.

Throws: IOException - if access to persistent storage fails.

SecurityException - if the Configuration object is bound to a location different from that of the calling bundle and it has no Admin-Permission.

public Configuration **getConfiguration**(java.lang.String pid,
    java.lang.String location)
    throws IOException, SecurityException

Get an existing Configuration object from the persistent store, or create a new Configuration object.

If a Configuration with this PID already exists in Configuration Admin service return it. The location parameter is ignored in this case.

Else, return a new Configuration object. This new object is bound to the location and the properties are set to null. If the location parameter is null, it will be set when a Managed Service with the corresponding PID is registered for the first time.

This method requires Admin Permission.

Parameters: pid - persistent identifier.

location - the bundle location string, or null.

Returns: an existing or new Configuration object.

Throws: IOException - if access to persistent storage fails.

SecurityException - if the caller does not have Admin Permission.

```
public Configuration[] listConfigurations(java.lang.String filter)
    throws IOException, InvalidSyntaxException
```

List the current Configuration objects which match the filter.

Only Configuration objects with non-null properties are considered current. That is, Configuration.getProperties() is guaranteed not to return null for each of the returned Configuration objects.

Normally only Configuration objects that are bound to the location of the calling bundle are returned. If the caller has Admin-Permission, then all matching Configuration objects are returned.

The syntax of the filter string is as defined in the Filter class. The filter can test any configuration parameters including the following system properties:

1. service.pid - String - the PID under which this is registered

2. service.factoryPid - String - the factory if applicable

3. service.bundleLocation - String - the bundle location

The filter can also be null, meaning that all Configuration objects should be returned.

Parameters: filter - a Filter object, or null to retrieve all Configuration objects.

Returns: all matching Configuration objects, or null if there aren't any

Throws: IOException - if access to persistent storage fails

InvalidSyntaxException - if the filter string is invalid

9.13.3 ConfigurationException

public class ConfigurationException extends java.lang.Exception

*All Implement-
ed Interfaces:*

java.io.Serializable

An Exception class to inform the Configuration Admin service of problems with configuration data.

9.13.3.1 Constructors

public **ConfigurationException**(java.lang.String property,
  java.lang.String reason)

Create a ConfigurationException object.

Parameters:

property - name of the property that caused the problem, null if no specific property was the cause

reason - reason for failure

9.13.3.2 Methods

public java.lang.String **getProperty**()

Return the property name that caused the failure or null.

public java.lang.String **getReason**()

Return the reason for this exception.

Returns:

reason of the failure

9.13.4 ConfigurationPlugin

public interface ConfigurationPlugin

A service interface for processing configuration dictionary before the update.

A bundle registers a ConfigurationPlugin object in order to process configuration updates before they reach the Managed Service or Managed Service Factory. The Configuration Admin service will detect registrations of Configuration Plugin services and must call these services every time before it calls the Managed Service or Managed Service Factory updated method. The Configuration Plugin service thus has the opportunity to view and modify the properties before they are passed to the ManagedS ervice or Managed Service Factory.

Configuration Plugin (plugin) services have full read/write access to all configuration information. Therefore, bundles using this facility should be trusted. Access to this facility should be limited with ServicePermission[REGISTER] for the Configuration Plugin service. Implementations of a Configuration Plugin service should assure that they only act on appropriate configurations.

The Integer service.cm Ranking registration property may be specified. Not specifying this registration property, or setting it to something other than an Integer, is the same as setting it to the Integer zero. The service.cm Ranking property determines the order in which plugins are invoked. Lower ranked plugins are called before higher ranked ones. In the event of more than one plugin having the same value of service.cm Ranking, then the Configuration Admin service arbitrarily chooses the order in which they are called.

By convention, plugins with service.cm Ranking< 0 or service.cm Ranking > 1000 should not make modifications to the properties.

The Configuration Admin service has the right to hide properties from plugins, or to ignore some or all the changes that they make. This might be done for security reasons. Any such behavior is entirely implementation defined.

A plugin may optionally specify a cm.target registration property whose value is the PID of the Managed Service or Managed Service Factory whose configuration updates the plugin is intended to intercept. The plugin will then only be called with configuration updates that are targetted at the Managed Service or Managed Service Factory with the specified PID. Omitting the cm.target registration property means that the plugin is called for all configuration updates.

9.13.4.1 Fields

public static final java.lang.String **CM_TARGET**

A service property to limit the Managed Service or Managed Service Factory configuration dictionaries a Configuration Plugin service receives. This property contains a String[] of PIDs. A Configuration Admin service must call a Configuration Plugin service only when this property is not set, or the target service's PID is listed in this property.

9.13.4.2

Methods

```
public void modifyConfiguration(ServiceReference reference,
    java.util.Dictionary properties)
```

View and possibly modify the a set of configuration properties before they are sent to the Managed Service or the Managed Service Factory. The Configuration Plugin services are called in increasing order of their service.cm Ranking property. If this property is undefined or is a non-Integer type, 0 is used.

This method should not modify the properties unless the service.cm Ranking of this plugin is in the range 0 <= service.cm Ranking <= 1000.

If this method throws any Exception, the Configuration Admin service must catch it and should log it.

Parameters: reference - reference to the Managed Service or Managed Service Factory

configuration - the configuration properties

9.13.5

ManagedService

public interface ManagedService

A service that can receive configuration data from a Configuration Admin service.

A Managed Service is a service that needs configuration data. Such an object should be registered with the Framework registry with the service.pid property set to some unique identitifier called a PID.

If the Configuration Admin service has a Configuration object corresponding to this PID, it will callback the updated() method of the Managed Service object, passing the properties of that Configuration object.

If it has no such Configuration object, then it calls back with a null properties argument. Registering a Managed Service will always result in a callback to the updated() method provided the Configuration Admin service is, or becomes active. This callback is always done on a different thread than the thread that performs the registration.

Else, every time that either of the updated() methods is called on that Configuration object, the Managed Service.updated() method with the new properties is called. If the delete() method is called on that Configuration object, Managed Service.updated() is called with a null for the properties parameter. All these callbacks are done on a different thread.

The following example shows the code of a serial port that will create a port depending on configuration information.

```java
class SerialPort implements ManagedService {
  ServiceRegistration registration;
  Hashtable configuration;
  CommPortIdentifier id;
  synchronized void open(CommPortIdentifier id,
  BundleContext context) {
    this.id = id;
    registration = context.registerService(
      ManagedService.class.getName(),
      this,
      getDefaults()
    );
  }
  Hashtable getDefaults() {
    Hashtable defaults = new Hashtable();
    defaults.put( "port", id.getName() );
    defaults.put( "product", "unknown" );
    defaults.put( "baud", "9600" );
    defaults.put( Constants.SERVICE_PID,
      "com.acme.serialport." + id.getName() );
    return defaults;
  }
  public synchronized void updated(
    Dictionary configuration  ) {
    if ( configuration == null )
      registration.setProperties( getDefaults() );
    else {
      setSpeed( configuration.get("baud") );
      registration.setProperties( configuration );
    }
  }
  ...
}
```

As a convention, it is recommended that when a Managed Service is updated, it should copy all the properties it does not recognize into the service registration properties. This will allow the Configuration Admin service to set properties on services which can then be used by other applications.

9.13.5.1

Methods

public void **updated**(java.util.Dictionary properties)
    throws ConfigurationException

Update the configuration for a Managed Service.

When the implementation of updated(Dictionary) detects any kind of error in the configuration properties, it should create a new ConfigurationException which describes the problem. This can allow a management system to provide useful information to a human administrator.

If this method throws any other Exception, the Configuration Admin service must catch it and should log it.

The Configuration Admin service must call this method on a thread other than the thread which initiated the callback. This implies that implementors of Managed Service can be assured that the callback will not take place during registration when they execute the registration in a synchronized method.

Parameters: properties - configuration properties, or null

Throws: ConfigurationException - when the update fails

9.13.6   ManagedServiceFactory

public interface ManagedServiceFactory

Manage multiple service instances. Bundles registering this interface are giving the Configuration Admin service the ability to create and configure a number of instances of a service that the implementing bundle can provide. For example, a bundle implementing a DHCP server could be instantiated multiple times for different interfaces using a factory.

Each of these *service instances* is represented, in the persistent storage of the Configuration Admin service, by a factory Configuration object that has a PID. When such a Configuration is updated, the Configuration Admin service calls the Managed Service Factory updated method with the new properties. When updated is called with a new PID, the Managed Service Factory should create a new factory instance based on these configuration properties. When called with a PID that it has seen before, it should update that existing service instance with the new configuration information.

In general it is expected that the implementation of this interface will maintain a data structure that maps PIDs to the factory instances that it has created. The semantics of a a factory instance are defined by the Managed Service Factory. However, if the factory instance is registered as a service object with the service registry, its PID should match the PID of the corresponding Configuration object.

An example that demonstrates the use of a factory. It will create serial ports under command of the Configuration Admin service.

```
class SerialPortFactory
  implements ManagedServiceFactory {
  ServiceRegistration registration;
  Hashtable ports;
  void start(BundleContext context) {
    Hashtable properties = new Hashtable();
    properties.put( Constants.SERVICE_PID,
      "com.acme.serialportfactory" );
    registration = context.registerService(
      ManagedService.class.getName(),
      this,
      properties
    );
  }
  public void updated( String pid,
    Dictionary properties  ) {
    String portName = (String) properties.get(
"port");
    SerialPortService port =
      (SerialPort) ports.get( pid );
    if ( port == null ) {
      port = new SerialPortService();
      ports.put( pid, port );
      port.open();
    }
    if ( port.getPortName().equals(portName) )
      return;
    port.setPortName( portName );
  }
  public void deleted( String pid ) {
    SerialPortService port =
      (SerialPort) ports.get( pid );
    port.close();
    ports.remove( pid );
  }
  ...
```

}

9.13.6.1 Methods

public void **deleted**(java.lang.String pid)

Remove a factory instance. Remove the factory instance associated with the PID. If the instance was registered with the service registry, it should be unregistered.

If this method throws any Exception, the Configuration Admin service must catch it and should log it.

The Configuration Admin service must call this method on a thread other than the thread which called delete() on the corresponding Configuration object.

Parameters: pid - the PID of the service to be removed

public java.lang.String **getName**()

Return a descriptive name of this factory.

Returns: the name for the factory, which might be localized

public void **updated**(java.lang.String pid, java.util.Dictionary
    properties)
    throws ConfigurationException

Create a new instance, or update the configuration of an existing instance. If the PID of the Configuration object is new for the Managed Service Factory, then create a new factory instance, using the configuration properties provided. Else, update the service instance with the provided properties.

If the factory instance is registered with the Framework, then the configuration properties should be copied to its registry properties. This is not mandatory and security sensitive properties should obviously not be copied.

If this method throws any Exception, the Configuration Admin service must catch it and should log it.

When the implementation of updated detects any kind of error in the configuration properties, it should create a new ConfigurationException which describes the problem.

The Configuration Admin service must call this method on a thread other than the thread which necessitated the callback. This implies that implementors of the ManagedServiceFactory class can be assured that the callback will not take place during registration when they execute the registration in a synchronized method.

Parameters: pid - the PID for this configuration

properties - the configuration properties

Throws: ConfigurationException - when the configuration properties are invalid

9.14   References

[26]   *DTMF Common Information Model*
http://www.dmtf.org

[27]   *Simple Network Management Protocol*
RFCs http://directory.google.com/Top/Computers/Internet/Protocols/SNMP/RFCs

[28]   *XSchema*
http://www.w3.org/TR/xmlschema-0/

[29]   *Interface Definition Language*
http://www.omg.org

[30]   *Lightweight Directory Access Protocol*
http://directory.google.com/Top/Computers/Software/Internet/Servers/Directory/LDAP

[31]   *Understanding and Deploying LDAP Directory services*
Timothy Howes et. al. ISBN 1-57870-070-1, MacMillan Technical publishing.

10 Metatype Specification

Version 1.0

10.1 Introduction

The Metatype specification defines interfaces that allow bundle developers to describe attribute types in a computer readable form using so-called *metadata*.

The purpose of this specification is to allow services to specify the type information of data that they can use as arguments. The data is based on *attributes*, which are key/value pairs like properties.

A designer in a type-safe language like Java is often confronted with the choice of using the language constructs to exchange data or using a technique based on attributes/properties that are based on key/value pairs. Attributes provide an escape from the rigid type-safety requirements of modern programming languages.

Type-safety works very well for software development environments in which multiple programmers work together on large applications or systems, but often lacks the flexibility needed to receive structured data from the outside world.

The attribute paradigm has several characteristics that make this approach suitable when data needs to be communicated between different entities which "speak" different languages. Attributes are uncomplicated, resilient to change, and allow the receiver to dynamically adapt to different types of data.

As an example, the OSGi Service Gateway Specification 1.0 defines several attribute types which are used in a Framework implementation, but which are also used and referenced by other OSGi specifications such as the *Configuration Admin Service Specification* on page 223. A Configuration Admin service implementation deploys attributes (key/value pairs) as configuration properties.

During the development of the Configuration Admin service, it became clear that the Framework attribute types needed to be described in a computer readable form. This information (the metadata) could then be used to automatically create user interfaces for management systems or could be translated into management information specifications such as CIM, SNMP and the like.

10.1.1 Essentials

- *Conceptual model* – The specification must have a conceptual model for how classes and attributes are organized.
- *Standards* – The specification should be aligned with appropriate standards, and explained in situations where the specification is not aligned with, or cannot be mapped to, standards.
- *Remote Management* – Remote management should be taken into account.
- *Size* – Minimal overhead in size for a bundle using this specification.
- *Localization* – It must be possible to use this specification with different languages at the same time. This ability allows servlets to serve information in the language selected in the browser.
- *Type information* – The definition of an attribution should contain the name (if it is required), the cardinality, a label, a description, labels for enumerated values, and the Java class that should be used for the values.
- *Validation* – It should be possible to validate the values of the attributes.

10.1.2 Operation

This specification starts with an object that implements the MetaTypeProvider interface. It is not specified how this object is obtained, and there are several possibilities. Often, however, this object is a service registered with the Framework.

A MetaTypeProvider object provides access to ObjectClassDefinition objects. These objects define all the information for a specific *object class*. An object class is a some descriptive information and a set of named attributes (which are key/value pairs).

Access to object classes is qualified by a locale and a Persistent IDentity (PID). The locale is a String object that defines for which language the ObjectClassDefinition is intended, allowing for localized user interfaces. The PID is used when a single MetaTypeProvider object can provide ObjectClassDefinition objects for multiple purposes. The context in which the MetaTypeProvider object is used should make this clear.

Attributes have global scope. Two object classes can consist of the same attributes, and attributes with the same name should have the same definition. This global scope is unlike many languages like Java that scope instance variables within a class, but it is similar to the Lightweight Directory Access Protocol (LDAP) and SNMP that also use a global attribute name space.

Attribute Definition objects provide sufficient localized information to generate user interfaces.

10.1.3 Entities

- *Attribute* – A key/value pair.
- *AttributeDefinition* – Defines a description, name, help text and type information of an attribute.
- *ObjectClassDefinition* – Defines the type of a datum. It contains a description and name of the type plus a set of AttributeDefinition objects.
- *MetaTypeProvider* – Provides access to the object classes that are available for this object. Access uses the PID and a locale to find the best ObjectClassDefinition object.

Figure 34   *Class Diagram Meta Typing, org.osgi.service.metatyping*

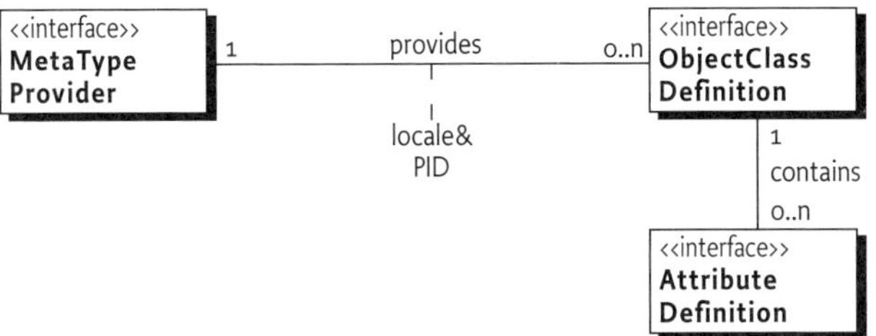

10.2 Attributes Model

The Framework uses the LDAP filter syntax for searching the Framework registry. The usage of the attributes in this specification and the Framework specification closely resemble the LDAP attribute model. Therefore, the names used in this specification have been aligned with LDAP. Consequently, the interfaces which are defined by this Specification are:

- AttributeDefinition
- ObjectClassDefinition
- MetaTypeProvider

These names correspond to the LDAP attribute model. For further information on ASN.1-defined attributes and X.500 object classes and attributes, see [33] *Understanding and Deploying LDAP Directory services.*

The LDAP attribute model assumes a global name space for attributes, and object classes consist of a number of attributes. So, if an object class inherits the same attribute from different parents, only one copy of the attribute must become part of the object class definition. This namespace implies that a given attribute, for example cn, should *always* be the common name and the type must always be a String. An attribute cn cannot be an Integer in another object class definition. In this respect, the OSGi approach towards attribute definitions is comparable with the LDAP attribute model.

10.3 Object Class Definition

The ObjectClassDefinition interface is used to group the attributes which are defined in AttributeDefinition objects.

An ObjectClassDefinition object contains the information about the overall set of attributes and has the following elements:

- A name which can be returned in different locales.
- A global namespace in the registry, which is the same condition as LDAP/X.500 object classes. In these standards the OSI Object Identifier (OID) is used to uniquely identify object classes. If such an OID exists, (which can be requested at several standard organizations, and many companies already have a node in the tree) it can be returned here. Otherwise, a unique id should be returned. This id can be a Java class name (reverse domain name) or generated with a GUID algorithm.
  All LDAP-defined object classes already have an associated OID. It is strongly advised to define the object classes from existing LDAP schemes which provide many preexisting OIDs. Many such schemes exist ranging from postal addresses to DHCP parameters.
- A human-readable description of the class.
- A list of attribute definitions which can be filtered as required or optional. Note that in X.500 the mandatory or required status of an attribute is part of the object class definition and not of the attribute definition.
- An icon, with an optional hint of its size.

10.4 Attribute Definition

The AttributeDefinition interface provides the means to describe the data type of attributes.

The AttributeDefinition interface defines the following elements:

- Defined names (final ints) for the data types as restricted in the Framework for the attributes, called the syntax in OSI terms, which can be obtained with the getType() method.
- AttributeDefinition objects should use a similar OID as described in the ID field for ObjectClassDefinition.
- A localized name intended to be used in user interfaces.

- A localized description that defines the semantics of the attribute and possible constraints, which should be usable for tooltips.
- An indication if this attribute should be stored as a unique value, a Vector, or an array of values. Also the maximum cardinality of the type.
- The data type, as limited by the Framework service registry attribute types.
- A validation function to verify if a possible value is correct.
- A list of values and a list of localized labels. Intended for popup menus in GUIs, allowing the user to choose from a set.
- A default value. The return type of this is a String[ ]. For cardinality = 0, this return type must be an array of 1 String. For other cardinalities, the array must not contain more than the absolute value of *cardinality* String objects. In that case, it may contain 0 objects.

10.5 Meta Type Provider

The MetaTypeProvider interface is used to access metatype information. It is used in management systems and runtime management. It supports locales so that the text used in AttributeDefinition and ObjectClassDefinition objects can be adapted to different locales.

The pid is given as an argument with the getObjectClassDefinition method so that a single MetaTypeProvider object can be used for different object classes with their own PID.

Locale objects are represented in String objects because not all profiles support Locale. The String holds the standard Locale presentation of:

```
<language> [ "_" <country> [ "_" <variation>]]
```

For example, "en", "nl_be", "en_ca_posix".

10.6 Metatype Example

AttributeDefinition and ObjectClassDefinition classes are intended to be easy to use for bundles. This example shows a naive implementation for these classes (note that the get methods usage are not shown). Commercial implementations can use XML, Java serialization, or Java Properties for implementations. This example uses plain code to store the definitions.

The example first shows that the ObjectClassDefinition interface is implemented in the OCD class. The name is made very short because the class is used to instantiate the static structures. Normally many of these objects are instantiated very close to each other, and long names would make these lists of instantiations very long.

```
class OCD implements ObjectClassDefinition {
    String                  name;
    String                  id;
    String                  description;
    AttributeDefinition     required[];
    AttributeDefinition     optional[];

    public OCD(
        String name, String id, String description,
        AttributeDefinition required[],
        AttributeDefinition optional[]) {

        this.name = name;
        this.id = id;
        this.description = description;
        this.required = required;
        this.optional = optional;
    }
    .... All the get methods
}
```

The second class is the AD class that implements the AttributeDefinition interface. The name is short for the same reason as in OCD. Note the two different constructors to simplify the common case.

```
class AD implements AttributeDefinition {
    String                  name;
    String                  id;
    String                  description;
    int                     cardinality;
    int                     syntax;
    String[]                values;
    String[]                labels;
    String[]                deflt;

    public AD( String name, String id, String
    description,
        int syntax, int cardinality, String values[],
        String labels[], String deflt[]) {
        this.name       = name;
        this.id         = id;
```

```
                    this.description    = description;
                    this.cardinality    = cardinality;
                    this.syntax         = syntax;
                    this.values         = values;
                    this.labels         = labels;
                }

            public AD( String name, String id, String
            description,
                int syntax)
            {
                this(name, id, description, syntax, 0, null,
            null, null);
                }
                ... All the get methods and validate method
            }
```

The last part is the example that implements a MetatTypeProvider
class. Only one locale is supported, the US locale. The OIDs used in
this example are the actual OIDs as defined in X.500.

```
        public class Example implements MetaTypeProvider {
            final static AD cn = new AD(
                "cn",          "2.5.4.3", "Common name", AD.STRING);
            final static AD sn = new AD(
                "sn",          "2.5.4.4", "Sur name", AD.STRING);
            final static AD description = new AD(
                "description", "2.5.4.13","Description",
        AD.STRING);
            final static AD seeAlso = new AD(
                "seeAlso",    "2.5.4.34", "See Also", AD.STRING);
            final static AD telephoneNumber = new AD(
                "telephoneNumber", "2.5.4.20", "Tel nr",
        AD.STRING);
            final static AD userPassword = new AD(
                "userPassword", "2.5.4.3", "Password",
        AD.STRING);

            final static ObjectClassDefinition person = new
        OCD(
                "person", "2.5.6.6", "Defines a person",
                  new AD[] { cn, sn },
                  new AD[] { description, seeAlso,
                    telephoneNumber, userPassword}
            );
```

```
    public ObjectClassDefinition
getObjectClassDefinition(
    String pid, String locale) {
    return person;
}

public String[] getLocales() {
    return new String[] { "en_US" };
}
}
```

This code shows that the attributes are defined in AD objects as final static. The example groups a number of attributes together in an OCD object.

As can be seen from this example, the resource issues for using AttributeDefinition, ObjectClassDefinition and MetaTypeProvider classes are minimized.

10.7 Related Standards

One of the primary goals of this specification is to make metatype information available at runtime with minimal overhead. Many related standards are applicable to metatypes; except for Java beans, however, all other metatype standards are based on document formats. In the OSGi environment, document format standards are deemed unsuitable due to the overhead required in the execution environment (they require a parser during runtime).

Another consideration is the applicability of these standards. Most of these standards were developed for management systems on platforms where resources are not necessarily a concern. In this case, a metatype standard is normally used to describe the data structures needed to control some other computer via a network. This other computer, however, does not require the metatype information as it is *implementing* this information.

In some traditional cases, a management system uses the metatype information to control objects in an OSGi environment. Therefore, the concepts and the syntax of the metatype information must be mappable to these popular standards. Clearly, then, these standards must be able to describe objects in an OSGi environment. This ability is usually not a problem, because the metatype languages used by current management systems are very powerful.

10.7.1 Beans

The intention of the Beans packages in Java comes very close to the metatype information needed in the OSGi environment. Beans packages cannot be used, however, for the following reasons:

- Beans packages require a large number of classes that are likely to be optional for an OSGi environment.
- Beans have been closely coupled to the graphic subsystem (AWT) and applets. Neither of these packages is available on an OSGi environment.
- Beans are closely coupled with the type-safe Java classes. The advantage of attributes is that no type-safety is used, allowing two parties to have an independent versioning model (no shared classes).
- Beans packages allow all possible Java objects, not the OSGi subset as required by this specification.
- Beans have no explicit localization.
- Beans have no support for optional attributes.

10.8 Security Considerations

Special security issues are not applicable for this specification.

10.9 org.osgi.service.metatype

The OSGi Metatype Package. Specification Version 1.0.

Bundles wishing to use this package must list the package in the Import-Package header of the bundle's manifest. For example:

```
Import-Package: org.osgi.service.metatype;
specification-version=1.0
```

Class Summary

Interfaces

AttributeDefinition	An interface to describe an attribute.
MetaTypeProvider	Provides access to metatypes.
ObjectClassDefinition	Description for the data type information of an objectclass.

10.9.1 AttributeDefinition

public interface AttributeDefinition

An interface to describe an attribute.

An AttributeDefinition object defines a description of the data type of a property/attribute.

10.9.1.1 Fields

public static final int **BIGDECIMAL**

The BIGDECIMAL(10) type. Attributes of this type should be stored as BigDecimal Vector with BigDecimal or BigDecimal[] objects depending on getCardinality().

public static final int **BIGINTEGER**

The BIGINTEGER(9) type. Attributes of this type should be stored as BigInteger, Vector with BigInteger or BigInteger[] objects, depending on the getCardinality() value.

public static final int **BOOLEAN**

The BOOLEAN(11) type. Attributes of this type should be stored as Boolean, Vector with Boolean or boolean[] objects depending on getCardinality().

public static final int **BYTE**

The BYTE(6) type. Attributes of this type should be stored as Byte, Vector with Byte or byte[] objects, depending on the getCardinality() value.

public static final int **CHARACTER**

The CHARACTER(5) type. Attributes of this type should be stored as Character, Vector with Character or char[] objects, depending on the getCardinality() value.

public static final int **DOUBLE**

The DOUBLE(7) type. Attributes of this type should be stored as Double, Vector with Double or double[] objects, depending on the getCardinality() value.

public static final int **FLOAT**

The FLOAT(8) type. Attributes of this type should be stored as Float Vector with Float or float[] objects, depending on the get-Cardinality() value.

public static final int **INTEGER**

The INTEGER(3) type. Attributes of this type should be stored as Integer, Vector with Integer or int[] objects, depending on the get-Cardinality() value.

public static final int **LONG**

The LONG(2) type. Attributes of this type should be stored as Long, Vector with Long or long[] objects, depending on the get-Cardinality() value.

public static final int **SHORT**

The SHORT(4) type. Attributes of this type should be stored as Short, Vector with Short or short[] objects, depending on the get-Cardinality() value.

public static final int **STRING**

The STRING(1) type.

Attributes of this type should be stored as String, Vector with String or String[] objects, depending on the get Cardinality() value.

10.9.1.2   **Methods**

public int **getCardinality**()

Return the cardinality of this attribute. The OSGi environment handles multi valued attributes in arrays ([]) or in Vector objects. The return value is defined as follows:

```
x = Integer.MIN_VALUE    no limit, but use Vector
 x < 0                   -x = max occurrences, store
in Vector
 x > 0                    x = max occurrences, store
in array []
 x = Integer.MAX_VALUE    no limit, but use array []
 x = 0                    1 occurrence required
```

public java.lang.String[] **getDefaultValue**()

Return a default for this attribute. The object must be of the appropriate type as defined by the cardinality and get Type(). The return type is a list of String objects that can be converted to the appropriate type. The cardinality of the return array must follow the absolute cardinality of this type. E.g. if the cardinality = 0, the array must contain 1 element. If the cardinality is 1, it must contain 0 or 1 elements. If it is -5, it must contain from 0 to max 5 elements. Note that the special case of a 0 cardinality, meaning a single value, does not allow arrays or vectors of 0 elements.

public java.lang.String **getDescription**()

Return a description of this attribute. The description may be localized and must describe the semantics of this type and any constraints.

public java.lang.String **getID**()

Unique identity for this attribute. Attributes share a global namespace in the registry. E.g. an attribute cn or common Name must always be a String and the semantics are always a name of some object. They share this aspect with LDAP/X.500 attributes. In these standards the OSI Object Identifier (OID) is used to uniquely identify an attribute. If such an OID exists, (which can be requested at several standard organisations and many companies already have a node in the tree) it can be returned here. Otherwise, a unique id should be returned which can be a Java class name (reverse domain name) or generated with a GUID algorithm. Note that all LDAP defined attributes already have an OID. It is strongly advised to define the attributes from existing LDAP schemes which will give the OID. Many such schemes exist ranging from postal addresses to DHCP parameters.

public java.lang.String **getName**()

Get the name of the attribute. This name may be localized.

public java.lang.String[] **getOptionLabels**()

Return a list of labels of option values.

The purpose of this method is to allow menus with localized labels. It is associated with get Option Values. The labels returned here are ordered in the same way as the values in that method.

If the function returns null, there are no option labels available.

This list must be in the same sequence as the getOptionValues() method. I.e. for each index i in getOptionLabels, i in getOptionValues() should be the associated value.

For example, if an attribute can have the value male, female, unknown, this list can return (for dutch) new String[] { "Man", "Vrouw", "Onbekend" }.

public java.lang.String[] **getOptionValues**()

Return a list of option values that this attribute can take.

If the function returns null, there are no option values available.

Each value must be acceptable to validate() (return "") and must be a String object that can be converted to the data type defined by getType() for this attribute.

This list must be in the same sequence as getOptionLabels(). I.e. for each index i in getOptionValues, i in getOptionLabels() should be the label.

For example, if an attribute can have the value male, female, unknown, this list can return new String[] { "male", "female", "unknown" }.

public int **getType**()

Return the type for this attribute.

Defined in the following constants which map to the appropriate Java type. STRING, LONG, INTEGER, CHAR, BYTE, DOUBLE, FLOAT, BIGINTEGER, BIGDECIMAL, BOOLEAN.

public java.lang.String **validate**(java.lang.String value)

Validate an attribute in String form. An attribute might be further constrained in value. This method will attempt to validate the attribute according to these constraints. It can return three different values:

```
null                    no validation present
" "                      no problems detected
"..."                   A localized description of why
the value is wrong
```

Parameters: value - The value before turning it into the basic data type

10.9.2   **MetaTypeProvider**

                    public interface MetaTypeProvider

10.9.2.1   **Methods**

                    public java.lang.String[] **getLocales**()

                    Return a list of locales available or null if only 1 The return parameter must be a name that consists of language [_ country [_ variation]] as is customary in the Locale class. This Locale class is not used because certain profiles do not contain it.

                    public ObjectClassDefinition **getObjectClassDefinition**(
                        java.lang.String pid, java.lang.String locale)

                    Return the definition of this object class for a locale.

                    The locale parameter must be a name that consists of language ["_" country ["_" variation]] as is customary in the Locale class. This Locale class is not used because certain profiles do not contain it.

                    The implementation should use the locale parameter to match an Object Class Definition object. It should follow the customary locale search path by removing the latter parts of the name.

Parameters: pid - The PID for which the type is needed or null if there is only 1

                    locale - The locale of the definition or null for default locale

10.9.3   **ObjectClassDefinition**

                    public interface ObjectClassDefinition

10.9.3.1   **Fields**

                    public static final int **ALL**

                    Argument for get Attribute Definitions(int). ALL indicates that all the definitions are returned. The value is -1.

                    public static final int **OPTIONAL**

                    Argument for get Attribute Definitions(int). OPTIONAL indicates that only the optional definitions are returned. The value is 2.

                    public static final int **REQUIRED**

                    Argument for get Attribute Definitions(int). REQUIRED indicates that only the required definitions are returned. The value is 1.

10.9.3.2   **Methods**

public AttributeDefinition[] **getAttributeDefinitions**(int filter)

Return the attribute definitions.

Return a set of attributes. The filter parameter can distinguish between ALL, REQUIRED or the OPTIONAL attributes.

Parameters: filter - ALL, REQUIRED, OPTIONAL

Returns: An array of attribute definitions or null if no attributes are selected

public java.lang.String **getDescription**()

Return a description of this object class. The description may be localized.

Returns: The localized description of the definition.

public java.io.InputStream **getIcon**(int size)

Return an InputStream object that can be used to create an icon from.

Indicate the size and return an InputStream object containing an icon. The returned icon maybe larger or smaller than the indicated size.

The icon may depend on the localization.

Parameters: sizeHint - size of an icon, e.g. a 16x16 pixels icon then size = 16

Returns: An InputStream representing an icon or null

Throws: IOException

public java.lang.String **getID**()

Return the id of this object class.

ObjectDefintion objects share a global namespace in the registry. They share this aspect with LDAP/X.500 attributes. In these standards the OSI Object Identifier (OID) is used to uniquely identify object classes. If such an OID exists, (which can be requested at several standard organisations and many companies already have a node in the tree) it can be returned here. Otherwise, a unique id should be returned which can be a java class name (reverse domain name) or generated with a GUID algorithm. Note that all LDAP defined object classes already have an OID associated. It is strongly advised to define the object classes from existing LDAP schemes which will give the OID for free. Many such schemes exist ranging from postal addresses to DHCP parameters.

Returns: The id or oid

public java.lang.String **getName**()

Return the name of this class.

Returns: The name of the described class.

10.10   References

[32]   *LDAP.*
       Available at http://directory.google.com/Top/Computers/Software/
       Internet/Servers/Directory/LDAP

[33]   *Understanding and Deploying LDAP Directory services*
       Timothy Howes et. al. ISBN 1-57870-070-1, MacMillan Technical
       publishing.

11   Preferences Service Specification

Version 1.0

11.1   Introduction

Many bundles need to save some data persistently. That is, the data
is required to survive the stopping and restarting of the bundle,
Framework and OSGi environment. In some cases, the data is specific to a particular user. For example, imagine a bundle that implements some kind of game. User specific persistent data could
include things like the user's preferred difficulty level for playing
the game. Some data is not specific to a user, which we call *system*
data. An example would be a table of high scores for the game.

Bundles which need to persist data in an OSGi environment can use
the file system via org.osgi.framework.BundleContext.getDataFile.
A file system, however, can only store bytes and characters, and provides no direct support for named values and different data types.

A popular class used to address this problem for Java applications is
the java.util.Properties class. This class allows data to be stored as
key/value pairs, called *properties*. For example, a property could have
a name "com.acme.fudd" and a value of "elmer". The Properties class
has rudimentary support for storage and retrieving with its load and
store methods. The Properties class, however, has the following
limitations:

- Does not support a naming hierarchy.
- Only supports String property values.
- Does not allow its content to be easily stored in a back-end
  system.
- Has no user namespace management.

Since the Properties class was introduced in Java 1.0, efforts have
been undertaken to replace it with a more sophisticated mechanism.
One of these efforts is this Preferences Service specification, which
in turn is aligned with the more extensive effort in [34] *JSR 10 Preferences API.*

11.1.1   ## Operation

The purpose of the Preferences Service specification is to allow bundles to store and retrieve properties stored in a tree of nodes, where each node implements the Preferences interface. The PreferencesService interface allows a bundle to create or obtain a Preferences tree for system properties, as well as a Preferences tree for each user of the bundle.

This specification allows for implementations where the data is stored locally on the service platform or remotely on a back-end system.

11.1.2   ## Essentials

The focus of this specification is simplicity, not reliable access to stored data. This specification does *not* define a general database service with transactions and atomicity guarantees. Instead, it is optimized to deliver the stored information when needed, but it will return defaults, instead of throwing an exception, when the back-end store is not available. This approach may reduce the reliability of the data, but it makes the service easier to use, and allows for a variety of compact and efficient implementations.

This API is made easier to use by the fact that many bundles can be written to ignore any problems that the Preferences Service may have in accessing the back-end store, if there is one. These bundles will mostly or exclusively use the methods of the Preferences interface which are not declared to throw a BackingStoreException.

This service only supports the storage of scalar values and byte arrays. It is not intended for storing large data objects like documents or images. No standard limits are placed on the size of data objects which can be stored, but implementations are expected to be optimized for the handling of small objects.

A hierarchical naming model is supported, in contrast to the flat model of the Properties class. A hierarchical model maps naturally to many computing problems. For example, maintaining information about the positions of adjustable chairs in a car requires information for each chair. In a hierarchy, this information can be modeled as a node per chair.

The Preferences interface API is closely aligned with [34] *JSR 10 Preferences API*. Code written for either of these APIs should require very little change when it is ported to the other.

A potential benefit of the Preferences Service is that it allows user specific preferences data to be kept in a well defined place, so that a user management system could locate it. This benefit could be useful for such operations as cleaning up files when a user is removed from the system, or to allow a user's preferences to be cloned for a new user.

The Preferences Service does *not* provide a mechanism to allow one bundle to access the preferences data of another. If a bundle wishes to allow another bundle to access its preferences data, it can pass a Preferences or PreferencesService object to that bundle.

The Preferences Service is not intended to provide configuration management functionality. For information regarding Configuration Management, refer to the *Configuration Admin Service Specification* on page 223.

11.1.3 Entities

The PreferencesService is a relatively simple service. It provides access to the different roots of Preferences trees. A single system root node and any number of user root nodes are supported. Each *node* of such a tree is an object that implements the Preferences interface.

This Preferences interface provides methods for traversing the tree, as well as methods for accessing the properties of the node. This interface also contains the methods to flush data into persistent storage, and to synchronize the in-memory data cache with the persistent storage.

All nodes except root nodes have a parent. Nodes can have multiple children.

Figure 35   *Preferences Class Diagram*

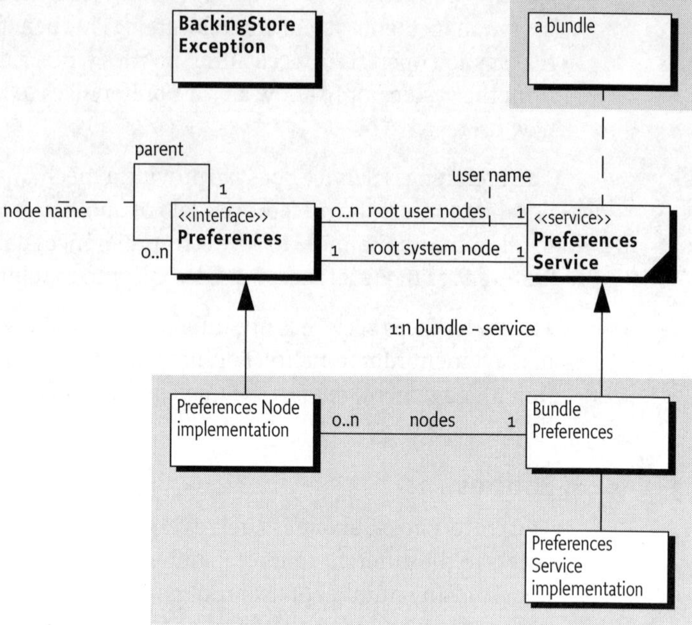

11.2   Preferences Interface

Preferences is an interface that defines the methods to manipulate a node and the tree to which it belongs. A Preferences object contains:

- A set of properties in the form of key/value pairs.
- A parent node.
- A number of child nodes.

11.2.1   Hierarchies

A valid Preferences object always belongs to a *tree*. A tree is identified by its root node. In such a tree, a Preferences object always has a single parent, except for a root node which has a null parent.

The root node of a tree can be found by recursively calling the parent() method of a node until null is returned. The nodes that are traversed this way are called the *ancestors* of a node.

Each Preferences object has a private name space for child nodes. Each child node has a name that must be unique among its siblings. Child nodes are created by getting a child node with the node(String) method. The String argument of this call contains a path name. Path names are explained in the next section.

Child nodes can have child nodes recursively. This group of objects is called the *descendants* of a node.

Descendants are automatically created when they are obtained from a Preferences object, including any intermediate nodes that are necessary for the given path. If this automatic creation is not desired, the nodeExists(String) method can be used to determine if a node already exists.

Figure 36   *Categorization of nodes in a tree*

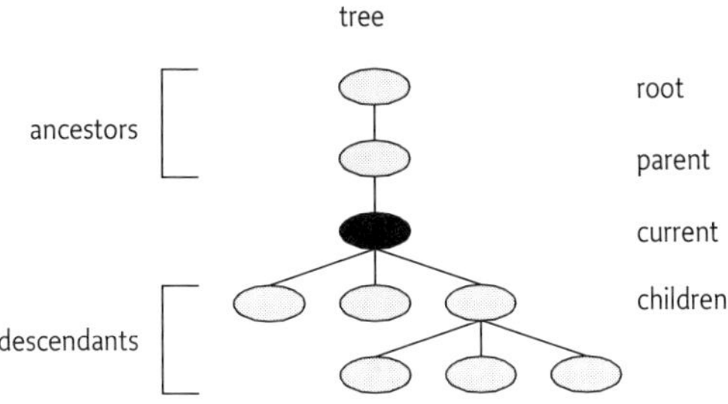

11.2.2   Naming

Each node has a name relative to its parent. A name may consist of Unicode characters except for the forward slash ("/"). There are no special names, like ".." or ".".

Empty names are reserved for root nodes. Node names that are directly created by a bundle must *always* contain at least one character.

Preferences node names and property keys are *case sensitive*: for example, "org.osgi" and "oRg.oSgl" are two distinct names. See *JSR 10* on page 295for the reasons why this differs from other OSGi services.

The Preferences Service supports different roots, so there is no absolute root for the Preferences Service. This concept is similar to [35] *Windows Registry* that also supports a number of roots.

A path consists of one or more node names, separated by a forward slash ("/"). Paths beginning with a "/" are called *absolute paths* while other paths are called *relative paths*. Paths cannot end with a "/" except for the special case of the root node which has absolute path "/".

Path names are always associated with a specific node; this node is called the current node in the following descriptions. Paths identify nodes as follows.

- *Absolute path* – The first "/" is removed from the path, and the remainder of the path is interpreted as a relative path from the tree's root node.
- *Relative path* –
  - If the path is the empty string, it identifies the current node.
  - If the path is a name (does not contain a "/"), then it identifies the child node with that name.
  - Otherwise, the first name from the path identifies a child of the current node. The name and slash are then removed from the path, and the remainder of the path is interpreted as a relative path from the child node.

11.2.3 Tree Traversal Methods

A tree can be traversed and modified with the following methods:

- childrenNames() – Returns the names of the child nodes.
- parent() – Returns the parent node.
- removeNode() – Removes this node and all its descendants.
- node(String) – Returns a Preferences object, which is created if it does not. already exist. The parameter is an absolute or relative path.
- nodeExists(String) – Returns true if the Preferences object identified by the path parameter exists.

11.2.4 Properties

Each Preferences node has a set of key/value pairs called properties. These properties consist of:

- *Key* – A key is a String object and *case sensitive.*
- The name space of these keys is separate from that of the child nodes. A Preferences node could have both a child node named "fudd" and a property named "fudd".
- *Value* – A value can always be stored and retrieved as a String object. Therefore, all primitive values must be encoded into String objects. A number of methods are available to store and retrieve values as primitive types. These methods are provided both for the convenience of the user of the Preferences interface, and to allow an implementation the option of storing the values in a more compact form.

All the keys that are defined in a Preferences object can be obtained with the keys() method. The clear() method can be used to clear all properties from a Preferences object. A single property can be removed with the remove(String) method.

11.2.5   ### Storing and Retrieving Properties

The Preferences interface has a number of methods for storing and retrieving property values based on their key. All the put* methods take as parameters a key and a value. All the get* methods take as parameters a key and a default value.

- put(String, String), get(String, String)
- putBoolean(String, boolean), getBoolean(String, boolean)
- putInt(String, int), getInt(String, int)
- putLong(String, long), getLong(String, long)
- putFloat(String, float), getFloat(String, float)
- putDouble(String, double), getDouble(String, double)
- putByteArray(String, byte[]), getByteArray(String, byte[])

The methods act as if all the values are stored as String objects, even though implementations may use different representations for the different types. For example, a property can be written as a String object and read back as a float, providing that the string can be parsed as a valid Java float. In the event of a parsing error, the get* methods do not raise exceptions, but instead return their default parameters.

11.2.6   ### Defaults

All get* methods take a default value as a parameter. The reasons for having such a default are:

- When a property for a Preferences object has not been set, the default is returned instead. In most cases, the bundle developer does not have to distinguish whether or not a property exists.
- A *best effort* strategy has been a specific design choice for this specification. The bundle developer should not have to react when the back-end store is not available. In those cases, the default value is returned without further notice.
  Bundle developers who want to assure that the back-end store is available should call the flush or sync method. Either of these methods will throw a BackingStoreException if the back-end store is not available.

11.3   Concurrency

This specification specifically allows an implementation to modify Preferences objects in a back-end store. If the back-end store is shared by multiple processes, concurrent updates may cause differences between the back-end store and the in-memory Preferences objects.

Bundle developers can partly control this concurrency with the flush() and sync() method. Both methods operate on a Preferences object.

The flush method performs the following actions:

- Stores (makes persistent) any ancestors (including the current node) that do not exist in the persistent store.
- Stores any properties which have been modified in this node since the last time it was flushed.
- Removes from the persistent store any child nodes that were removed from this object since the last time it was flushed.
- Flushes all existing child nodes.

The sync method will first flush, and then ensure that any changes that have been made to the current node and its descendents in the back-end store (by some other process) take effect. For example, it could fetch all the descendants into a local cache, or it could clear all the descendants from the cache so that they will be read from the back-end store as required.

If either method fails, a BackingStoreException is thrown.

The flush or sync methods provide no atomicity guarantee. When updates to the same back-end store are done concurrently by two different processes, the result may be that changes made by different processes are intermingled. To avoid this problem, implementations may simply provide a dedicated section (or namespace) in the back-end store for each OSGi environment, so that clashes do not arise, in which case there is no reason for bundle programmers to ever call sync.

In cases where sync is used, the bundle programmer needs to take into account that changes from different processes may become intermingled, and the level of granularity that can be assumed is the individual property level. Hence, for example, if two properties need to be kept in lockstep, so that one should not be changed without a corresponding change to the other, consider combining them into a single property, which would then need to be parsed into its two constituent parts.

11.4 PreferencesService Interface

The PreferencesService is obtained from the Framework's service registry in the normal way. Its purpose is to provide access to Preferences root nodes.

A Preferences Service maintains a system root and a number of user roots. User roots are automatically created, if necessary, when they are requested. Roots are maintained on a per bundle basis. For example, a user root called "elmer" in one bundle is a distinct from a user root with the same name in another bundle. Also, each bundle has its own system root. Implementations should use a ServiceFactory service object to create a separate PreferencesService object for each bundle.

The precise description of *user* and *system* will vary from one bundle to another. The Preference Service only provides a mechanism, the bundle may use this mechanism in any desired way.

The PreferencesService interface has the following methods to access the system root and user roots:

- getSystemPreferences() – Return a Preferences object that is the root of the system preferences tree.
- getUserPreferences(String) – Return a Preferences object associated with the user name that is given as argument. If the user does not exist, a new root is created atomically.
- getUsers() – Return an array of the names of all the users for whom a Preferences tree exists.

11.5 Cleanup

The Preferences Service must listen for bundle uninstall events, and remove all the preferences data for the bundle that is being uninstalled.

It also must handle the possibility of a bundle getting uninstalled while the Preferences Service is stopped. Therefore, it must check on startup whether preferences data exists for any bundle which is not currently installed. If it does, that data must be removed.

11.6 JSR 10

The Preferences specification is aligned with the Java Community Process JSR 10. This JSR defines an API that is intended to become part of a future Java Standard Edition.

The following differences between this specification and JSR 10 apply (from the point of view of this specification):

- Different package name.
- Preferences is an interface instead of a class.
- No listener facility is available to detect changes in a Preferences object.

- No maximum value is defined for the storage length, nor for the length of the key. JSR 10 has the constants MAX_KEY_LENGTH, MAX_VALUE_LENGTH, MAX_NAME_LENGTH.
- No factory methods for creation of Preferences objects. Objects are obtained from the Preferences Service.
- No isUserNode method.

11.7 org.osgi.service.prefs

The OSGi Preferences Service Package. Specification Version 1.0.

Bundles wishing to use this package must list the package in the Import-Package header of the bundle's manifest. For example:

```
Import-Package: org.osgi.service.prefs;
specification-version=1.0
```

Class Summary

Interfaces

Preferences	A node in a hierarchical collection of preference data.
PreferencesService	The Preferences Service.

Exceptions

BackingStoreException	Thrown to indicate that a preferences operation could not complete because of a failure in the backing store, or a failure to contact the backing store.

11.7.1 BackingStoreException

public class BackingStoreException extends java.lang.Exception

All Implemented Interfaces: java.io.Serializable

Thrown to indicate that a preferences operation could not complete because of a failure in the backing store, or a failure to contact the backing store.

11.7.1.1 Constructors

public **BackingStoreException**(java.lang.String s)

Constructs a BackingStoreException with the specified detail message.

11.7.2   ## Preferences

public interface Preferences

A node in a hierarchical collection of preference data.

This interface allows applications to store and retrieve user and system preference data. This data is stored persistently in an implementation-dependent backing store. Typical implementations include flat files, OS-specific registries, directory servers and SQL databases.

For each bundle, there is a separate tree of nodes for each user, and one for system preferences. The precise description of "user" and "system" will vary from one bundle to another. Typical information stored in the user preference tree might include font choice, and color choice for a bundle which interacts with the user via a servlet. Typical information stored in the system preference tree might include installation configuration data, or things like high score information for a game program.

Nodes in a preference tree are named in a similar fashion to directories in a hierarchical file system. Every node in a preference tree has a *node name* (which is not necessarily unique), a unique *absolute path name*, and a path name *relative* to each ancestor including itself.

The root node has a node name of the empty String object (""). Every other node has an arbitrary node name, specified at the time it is created. The only restrictions on this name are that it cannot be the empty string, and it cannot contain the slash character ('/').

The root node has an absolute path name of "/". Children of the root node have absolute path names of "/" + *⟨node name⟩*. All other nodes have absolute path names of *⟨parent's absolute path name⟩* + "/" + *⟨node name⟩*. Note that all absolute path names begin with the slash character.

A node *n*'s path name relative to its ancestor *a* is simply the string that must be appended to *a*'s absolute path name in order to form *n*'s absolute path name, with the initial slash character (if present) removed. Note that:

- No relative path names begin with the slash character.
- Every node's path name relative to itself is the empty string.
- Every node's path name relative to its parent is its node name (except for the root node, which does not have a parent).
- Every node's path name relative to the root is its absolute path name with the initial slash character removed.

Note finally that:

- No path name contains multiple consecutive slash characters.
- No path name with the exception of the root's absolute path name end in the slash character.
- Any string that conforms to these two rules is a valid path name.

Each Preference node has zero or more properties associated with it, where a property consists of a name and a value. The bundle writer is free to choose any appropriate names for properties. Their values can be of type String, long, int, boolean byte[], float, or double but they can always be accessed as if they were String objects.

All node name and property name comparisons are case-sensitive!

All of the methods that modify preference data are permitted to operate asynchronously; they may return immediately, and changes will eventually propagate to the persistent backing store, with an implementation-dependent delay. The flush method may be used to synchronously force updates to the backing store.

Implementations must automatically attempt to flush to the backing store any pending updates for a bundle's preferences when the bundle is stopped or otherwise ungets the Preferences Service.

The methods in this class may be invoked concurrently by multiple threads in a single Java Virtual Machine (JVM) without the need for external synchronization, and the results will be equivalent to some serial execution. If this class is used concurrently *by multiple JVMs* that store their preference data in the same backing store, the data store will not be corrupted, but no other guarantees are made concerning the consistency of the preference data.

11.7.2.1   **Methods**

public java.lang.String **absolutePath**()

Returns this node's absolute path name. Note that:

- The path name of the root node is "/".
- Path names other than that of the root node may not end in slash ('/').
- "." and ".." have *no* special significance in path names.
- The only illegal path names are those that contain multiple consecutive slashes, or that end in slash and are not the root.

Returns: this node's absolute path name.

public java.lang.String[] **childrenNames**()
    throws BackingStoreException, IllegalStateException

Returns the names of the children of this node. (The returned array will be of size zero if this node has no children and not null!)

Returns: the names of the children of this node.

Throws: BackingStoreException - if this operation cannot be completed due
                       to a failure in the backing store, or inability to communicate with it.

                       IllegalStateException - if this node (or an ancestor) has been
                       removed with the removeNode() method.

public void **clear**()
    throws BackingStoreException, IllegalStateException

Removes all of the properties (key-value associations) in this node.
This call has no effect on any descendants of this node.

Throws: BackingStoreException - if this operation cannot be completed due
                       to a failure in the backing store, or inability to communicate with it.

                       IllegalStateException - if this node (or an ancestor) has been
                       removed with the removeNode() method.

See Also: remove(String)

public void **flush**()
    throws BackingStoreException, IllegalStateException

Forces any changes in the contents of this node and its descendants
to the persistent store.

Once this method returns successfully, it is safe to assume that all
changes made in the subtree rooted at this node prior to the method
invocation have become permanent.

Implementations are free to flush changes into the persistent store
at any time. They do not need to wait for this method to be called.

When a flush occurs on a newly created node, it is made persistent,
as are any ancestors (and descendants) that have yet to be made persistent. Note however that any properties value changes in ancestors
are *not* guaranteed to be made persistent.

Throws: BackingStoreException - if this operation cannot be completed due
                       to a failure in the backing store, or inability to communicate with it.

                       IllegalStateException - if this node (or an ancestor) has been
                       removed with the removeNode() method.

See Also: sync()

public java.lang.String **get**(java.lang.String key, java.lang.String def)
    throws IllegalStateException, NullPointerException

Returns the value associated with the specified key in this node. Returns the specified default if there is no value associated with the key, or the backing store is inaccessible.

Parameters: key - key whose associated value is to be returned.

def - the value to be returned in the event that this node has no value associated with key or the backing store is inaccessible.

Returns: the value associated with key, or def if no value is associated with key.

Throws: IllegalStateException - if this node (or an ancestor) has been removed with the removeNode() method.

NullPointerException - if key is null. (A null default *is* permitted.)

public boolean **getBoolean**(java.lang.String key, boolean def)
    throws NullPointerException, IllegalStateException

Returns the boolean value represented by the String object associated with the specified key in this node. Valid strings are "true", which represents true, and "false", which represents false. Case is ignored, so, for example, "TRUE" and "False" are also valid. This method is intended for use in conjunction with the putBoolean(String, boolean) method.

Returns the specified default if there is no value associated with the key, the backing store is inaccessible, or if the associated value is something other than "true" or "false", ignoring case.

Parameters: key - key whose associated value is to be returned as a boolean.

def - the value to be returned in the event that this node has no value associated with key or the associated value cannot be interpreted as a boolean or the backing store is inaccessible.

Returns: the boolean value represented by the String object associated with key in this node, or null if the associated value does not exist or cannot be interpreted as a boolean.

Throws: NullPointerException - if key is null.

IllegalStateException - if this node (or an ancestor) has been removed with the removeNode() method.

See Also: get(String, String), putBoolean(String, boolean)

public byte[] **getByteArray**(java.lang.String key, byte[] def)
    throws NullPointerException, IllegalStateException

Returns the byte[] value represented by the String object associated with the specified key in this node. Valid String objects are *Base64* encoded binary data, as defined in RFC 2045, Section 6.8, with one minor change: the string must consist solely of characters from the *Base64 Alphabet*; no newline characters or extraneous characters are permitted. This method is intended for use in conjunction with the putByteArray(String, byte[]) method.

Returns the specified default if there is no value associated with the key, the backing store is inaccessible, or if the associated value is not a valid Base64 encoded byte array (as defined above).

Parameters: key - key whose associated value is to be returned as a byte[] object.

def - the value to be returned in the event that this node has no value associated with key or the associated value cannot be interpreted as a byte[] type, or the backing store is inaccessible.

Returns: the byte[] value represented by the String object associated with key in this node, or def if the associated value does not exist or cannot be interpreted as a byte[].

Throws: NullPointerException - if key is null.

IllegalStateException - if this node (or an ancestor) has been removed with the removeNode() method.

See Also: get(String, String), putByteArray(String, byte[])

public double **getDouble**(java.lang.String key, double def)
    throws IllegalStateException, NullPointerException

Returns the double value represented by the String object associated with the specified key in this node. The String object is converted to an int value as by Double.parseDouble(String). Returns the specified default if there is no value associated with the key, the backing store is inaccessible, or if Double.parseDouble(String) would throw a NumberFormatException if the associated value were passed. This method is intended for use in conjunction with the putDouble(String, double) method.

Parameters: key - key whose associated value is to be returned as a double value.

def - the value to be returned in the event that this node has no value associated with key or the associated value cannot be interpreted as a double type or the backing store is inaccessible.

Returns: the double value represented by the String object associated with key in this node, or def if the associated value does not exist or cannot be interpreted as a double type.

Throws: IllegalStateException - if this node (or an ancestor) has been removed with the the removeNode() method.

NullPointerException - if key is null.

See Also: putDouble(String, double), get(String, String)

public float **getFloat**(java.lang.String key, float def)
    throws IllegalStateException, NullPointerException

Returns the float value represented by the String object associated with the specified key in this node. The String object is converted to an int value as by Float.parseFloat(String). Returns the specified default if there is no value associated with the key, the backing store is inaccessible, or if Float.parseFloat(String) would throw a NumberFormatException if the associated value were passed. This method is intended for use in conjunction with the putFloat(String, float) method.

Parameters: key - key whose associated value is to be returned as a float value.

def - the value to be returned in the event that this node has no value associated with key or the associated value cannot be interpreted as a float type or the backing store is inaccessible.

Returns: the float value represented by the string associated with key in this node, or def if the associated value does not exist or cannot be interpreted as a float type.

Throws: IllegalStateException - if this node (or an ancestor) has been removed with the removeNode() method.

NullPointerException - if key is null.

See Also: putFloat(String, float), get(String, String)

public int **getInt**(java.lang.String key, int def)
    throws NullPointerException, IllegalStateException

Returns the int value represented by the String object associated with the specified key in this node. The String object is converted to an int as by Integer.parseInt(String). Returns the specified default if there is no value associated with the key, the backing store is inaccessible, or if Integer.parseInt(String) would throw a NumberFormatException if the associated value were passed. This method is intended for use in conjunction with the putInt(String, int) method.

Parameters:	key - key whose associated value is to be returned as an int.
	def - the value to be returned in the event that this node has no value associated with key or the associated value cannot be interpreted as an int or the backing store is inaccessible.
Returns:	the int value represented by the String object associated with key in this node, or def if the associated value does not exist or cannot be interpreted as an int type.
Throws:	NullPointerException - if key is null.
	IllegalStateException - if this node (or an ancestor) has been removed with the removeNode() method.
See Also:	putInt(String, int), get(String, String)

public long **getLong**(java.lang.String key, long def)
    throws NullPointerException, IllegalStateException

Returns the long value represented by the String object associated with the specified key in this node. The String object is converted to a long as by Long.parseLong(String). Returns the specified default if there is no value associated with the key, the backing store is inaccessible, or if Long.parseLong(String) would throw a Number-FormatException if the associated value were passed. This method is intended for use in conjunction with the putLong(String, long) method.

Parameters:	key - key whose associated value is to be returned as a long value.
	def - the value to be returned in the event that this node has no value associated with key or the associated value cannot be interpreted as a long type or the backing store is inaccessible.
Returns:	the long value represented by the String object associated with key in this node, or def if the associated value does not exist or cannot be interpreted as a long type.
Throws:	NullPointerException - if key is null.
	IllegalStateException - if this node (or an ancestor) has been removed with the removeNode() method.
See Also:	putLong(String, long), get(String, String)

public java.lang.String[] **keys**()
    throws BackingStoreException, IllegalStateException

Returns all of the keys that have an associated value in this node. (The returned array will be of size zero if this node has no preferences and not null!)

Returns: an array of the keys that have an associated value in this node.

Throws: BackingStoreException - if this operation cannot be completed due to a failure in the backing store, or inability to communicate with it.

                      IllegalStateException - if this node (or an ancestor) has been removed with the removeNode() method.

public java.lang.String **name**()

Returns this node's name, relative to its parent.

Returns: this node's name, relative to its parent.

public Preferences **node**(java.lang.String pathName)
    throws IllegalArgumentException, IllegalStateException

Returns a named Preferences object (node), creating it and any of its ancestors if they do not already exist. Accepts a relative or absolute pathname. Absolute pathnames (which begin with '/') are interpreted relative to the root of this node. Relative pathnames (which begin with any character other than '/') are interpreted relative to this node itself. The empty string ("") is a valid relative pathname, referring to this node itself.

If the returned node did not exist prior to this call, this node and any ancestors that were created by this call are not guaranteed to become persistent until the flush method is called on the returned node (or one of its descendants).

Parameters: pathName - the path name of the Preferences object to return.

Returns: the specified Preferences object.

Throws: IllegalArgumentException - if the path name is invalid.

                      IllegalStateException - if this node (or an ancestor) has been removed with the removeNode() method.

See Also: flush()

public boolean **nodeExists**(java.lang.String pathName)
    throws BackingStoreException, IllegalStateException

Returns true if the named node exists. Accepts a relative or absolute pathname. Absolute pathnames (which begin with '/') are interpreted relative to the root of this node. Relative pathnames (which begin with any character other than '/') are interpreted relative to this node itself. The pathname "" is valid, and refers to this node itself.

If this node (or an ancestor) has already been removed with the removeNode() method, it *is* legal to invoke this method, but only with the pathname ""; the invocation will return false. Thus, the idiom p.nodeExists("") may be used to test whether p has been removed.

Parameters: pathName - the path name of the node whose existence is to be checked.

Returns: true if the specified node exists.

Throws: BackingStoreException - if this operation cannot be completed due to a failure in the backing store, or inability to communicate with it.

IllegalStateException - if this node (or an ancestor) has been removed with the removeNode() method and pathname is not the empty string ("").

public Preferences **parent**()
    throws IllegalStateException

Returns the parent of this node, or null if this is the root.

Returns: the parent of this node.

Throws: IllegalStateException - if this node (or an ancestor) has been removed with the removeNode() method.

public void **put**(java.lang.String key, java.lang.String value)
    throws NullPointerException, IllegalStateException

Associates the specified value with the specified key in this node.

Parameters: key - key with which the specified value is to be associated.

value - value to be associated with the specified key.

Throws: NullPointerException - if key or value is null.

IllegalStateException - if this node (or an ancestor) has been removed with the removeNode() method.

public void **putBoolean**(java.lang.String key, boolean value)
    throws NullPointerException, IllegalStateException

Associates a String object representing the specified boolean value with the specified key in this node. The associated string is "true" if the value is true, and "false" if it is false. This method is intended for use in conjunction with the getBoolean(String, boolean) method.

Implementor's note: it is *not* necessary that the value be represented by a string in the backing store. If the backing store supports boolean values, it is not unreasonable to use them. This implementation detail is not visible through the Preferences API, which allows the value to be read as a boolean (with getBoolean) or a String (with get) type.

Parameters: key - key with which the string form of value is to be associated.

value - value whose string form is to be associated with key.

Throws: NullPointerException - if key is null.

IllegalStateException - if this node (or an ancestor) has been removed with the removeNode() method.

See Also: getBoolean(String, boolean), get(String, String)

public void **putByteArray**(java.lang.String key, byte[] value)
    throws NullPointerException, IllegalStateException

Associates a String object representing the specified byte[] with the specified key in this node. The associated String object the *Base64* encoding of the byte[], as defined in RFC 2045, Section 6.8, with one minor change: the string will consist solely of characters from the *Base64 Alphabet*; it will not contain any newline characters. This method is intended for use in conjunction with the getByteArray(String, byte[]) method.

Implementor's note: it is *not* necessary that the value be represented by a String type in the backing store. If the backing store supports byte[] values, it is not unreasonable to use them. This implementation detail is not visible through the Preferences API, which allows the value to be read as an a byte[] object (with getByteArray) or a String object (with get).

Parameters: key - key with which the string form of value is to be associated.

value - value whose string form is to be associated with key.

Throws: NullPointerException - if key or value is null.

IllegalStateException - if this node (or an ancestor) has been removed with the removeNode() method.

See Also: getByteArray(String, byte[]), get(String, String)

public void **putDouble**(java.lang.String key, double value)
    throws NullPointerException, IllegalStateException

Associates a String object representing the specified double value with the specified key in this node. The associated String object is the one that would be returned if the double value were passed to Double.to String(double). This method is intended for use in conjunction with the getDouble(String, double) method

Implementor's note: it is *not* necessary that the value be represented by a string in the backing store. If the backing store supports double values, it is not unreasonable to use them. This implementation detail is not visible through the Preferences API, which allows the value to be read as a double (with get Double) or a String (with get) type.

Parameters: key - key with which the string form of value is to be associated.

value - value whose string form is to be associated with key.

Throws: NullPointerException - if key is null.

IllegalStateException - if this node (or an ancestor) has been removed with the removeNode() method.

See Also: getDouble(String, double)

public void **putFloat**(java.lang.String key, float value)
    throws NullPointerException, IllegalStateException

Associates a String object representing the specified float value with the specified key in this node. The associated String object is the one that would be returned if the float value were passed to Float.to String(float). This method is intended for use in conjunction with the getFloat(String, float) method.

Implementor's note: it is *not* necessary that the value be represented by a string in the backing store. If the backing store supports float values, it is not unreasonable to use them. This implementation detail is not visible through the Preferences API, which allows the value to be read as a float (with get Float) or a String (with get) type.

Parameters: key - key with which the string form of value is to be associated.

value - value whose string form is to be associated with key.

Throws: NullPointerException - if key is null.

IllegalStateException - if this node (or an ancestor) has been removed with the removeNode() method.

See Also: getFloat(String, float)

```
public void putInt(java.lang.String key, int value)
    throws NullPointerException, IllegalStateException
```

Associates a String object representing the specified int value with the specified key in this node. The associated string is the one that would be returned if the int value were passed to Integer.to String(int). This method is intended for use in conjunction with getInt(String, int) method.

Implementor's note: it is *not* necessary that the property value be represented by a String object in the backing store. If the backing store supports integer values, it is not unreasonable to use them. This implementation detail is not visible through the Preferences API, which allows the value to be read as an int (with getInt or a String (with get) type.

Parameters: key - key with which the string form of value is to be associated.

                   value - value whose string form is to be associated with key.

Throws: NullPointerException - if key is null.

                   IllegalStateException - if this node (or an ancestor) has been removed with the removeNode() method.

See Also: getInt(String, int)


```
public void putLong(java.lang.String key, long value)
    throws NullPointerException, IllegalStateException
```

Associates a String object representing the specified long value with the specified key in this node. The associated String object is the one that would be returned if the long value were passed to Long.to String(long). This method is intended for use in conjunction with the getLong(String, long) method.

Implementor's note: it is *not* necessary that the value be represented by a String type in the backing store. If the backing store supports long values, it is not unreasonable to use them. This implementation detail is not visible through the Preferences API, which allows the value to be read as a long (with getLong or a String (with get) type.

Parameters: key - key with which the string form of value is to be associated.

                   value - value whose string form is to be associated with key.

Throws: NullPointerException - if key is null.

                   IllegalStateException - if this node (or an ancestor) has been removed with the removeNode() method.

See Also: getLong(String, long)

public void **remove**(java.lang.String key)
    throws IllegalStateException

Removes the value associated with the specified key in this node, if any.

Parameters: key - key whose mapping is to be removed from this node.

Throws: IllegalStateException - if this node (or an ancestor) has been removed with the removeNode() method.

See Also: get(String, String)

public void **removeNode**()
    throws IllegalStateException, RuntimeException

Removes this node and all of its descendants, invalidating any properties contained in the removed nodes. Once a node has been removed, attempting any method other than name(), absolutePath( ) or nodeExists("") on the corresponding Preferences instance will fail with an IllegalStateException. (The methods defined on Object can still be invoked on a node after it has been removed; they will not throw IllegalStateException.)

The removal is not guaranteed to be persistent until the flush method is called on the parent of this node. (It is illegal to remove the root node.)

Throws: IllegalStateException - if this node (or an ancestor) has already been removed with the removeNode() method.

RuntimeException - if this is a root node.

BackingStoreException

See Also: flush()

public void **sync**()
    throws BackingStoreException, IllegalStateException

Ensures that future reads from this node and its descendants reflect any changes that were committed to the persistent store (from any VM) prior to the sync invocation. As a side-effect, forces any changes in the contents of this node and its descendants to the persistent store, as if the flush method had been invoked on this node.

Throws: BackingStoreException - if this operation cannot be completed due to a failure in the backing store, or inability to communicate with it.

IllegalStateException - if this node (or an ancestor) has been removed with the removeNode() method.

See Also: flush()

11.7.3   **PreferencesService**

public interface PreferencesService

The Preferences Service.

Each bundle using this service has its own set of preference trees: one for system preferences, and one for each user.

A Preferences Service object is specific to the bundle which obtained it from the service registry. If a bundle wishes to allow another bundle to access its preferences, it should pass its Preferences Service object to that bundle.

11.7.3.1   **Methods**

public Preferences **getSystemPreferences**()

Returns the root system node for the calling bundle.

public Preferences **getUserPreferences**(java.lang.String name)

Returns the root node for the specified user and the calling bundle.

public java.lang.String[] **getUsers**()

Returns the names of users for which node trees exist.

11.8   References

[34] *JSR 10 Preferences API*
      http://jcp.org/aboutJava/communityprocess/review/jsr010/
      index.html

[35] *Windows Registry*
      http://www.microsoft.com/technet/win98/reg.asp

[36] *RFC 2045 Base 64 encoding*
      http://www.ietf.org/rfc/rfc2045.txt

12 User Admin Service Specification

Version 1.0

12.1 Introduction

OSGi environments are often be used in places where end users or devices initiate actions. These kinds of actions inevitably create a need for authenticating the initiator. Authenticating can be done in many different ways: with passwords, one-time token cards, biometrics, and certificates.

Once the initiator is authenticated, it is necessary to verify that this principal is authorized to perform the requested action. This authorization can only be decided by the operator of the OSGi environment, and thus requires administration.

The User Admin service provides this type of functionality. Bundles can use the User Admin service to authenticate an initiator and represent this authentication as an Authorization object. Bundles that execute actions on behalf of this user can use the Authorization object to verify if that user is authorized.

The User Admin service provides authorization based on who runs the code, instead of the Java code-based permission model, see [37] *The Java Security Architecture for JDK 1.2*. It is performing a role similar to [38] *Java Authentication and Authorization Service.*

12.1.1 Essentials

- *Authentication* – A large number of authentication schemes already exist, and more will come. The User Admin service must be flexible enough to adapt to all schemes that can be run on a computer system.
- *Authorization* – All bundles should use the User Admin service to authenticate users and to find out if those users are authorized. It is therefore paramount that a bundle can find out the authorization information with little effort.
- *Security* – Fine-grained security based on the Framework security model is needed to make access to the User Admin service safe. It should allow limited access to the credentials and other properties.

- *Extensibility* – Other bundles should be able to build on the User Admin service. It should therefore be possible to examine the information of this service and get real-time notifications of changes.
- *Properties* – The User Admin service must hold a persistent database of users. It must be possible to use this database to hold more information about this user.
- *Administration* – Administering the authorizations for each possible action and initiator is time-consuming and error-prone. It is therefore necessary to have mechanisms to group end users and to make it simple to assign authorizations to all members of a group at once.

12.1.2 Operation

An operator uses the User Admin service to define OSGi environment users and configure them with properties, credentials, and *roles.*

A Role object represents the initiator of a request (human or otherwise). This specification defines two types of roles:

- *User* – A User object can be configured with credentials (such as a password) and properties (for example, address, telephone number, and so on).
- *Group* – A Group object is an aggregation of *basic* and *required* roles. Basic and required roles are used in the authorization phase.

An OSGi environment may have several entry points, each of which will be responsible for authenticating incoming requests. An example of an entry point is the Http Service, which delegates authentication of incoming requests to the handleSecurity method of the HttpContext object that was specified when the target servlet or resource of the request was registered.

The OSGi environment entry points should use the information configured in the User Admin service to authenticate incoming requests: for example, a password stored in the private credentials, or the use of a certificate.

A bundle can determine if a request for an action is authorized by looking for a Role object that has the name of the requested action.

The bundle may execute the action if the Role object representing the initiator *implies* the Role object representing the requested action.

An initiator Role object *I* implies an action Group object *A* if:

- *I* implies at least one of *A*'s basic members, and

- *I* implies all of *A*'s required members.

An initiator Role object *I* implies an action User object *A* if:

- *A* and *I* are equal.

The Authorization class handles this non-trivial logic. The User Admin service can capture the privileges of an authenticated User object into an Authorization object. The Authorization.hasRole method checks if the authenticate User object has (or implies) a specified action Role object.

For example, in the case of the Http Service, the HttpContext object can authenticate the initiator and place an Authorization object in the request header. The servlet calls the hasRole method on this Authorization object to verify that the initiator has the authority to perform a certain action.

12.1.3 Entities

This Specification defines the following User Admin service entities:

- *UserAdmin* – This interface manages a database of named roles which can be used for authorization and authentication purposes.
- *Role* – This interface exposes the characteristics shared by all roles: a name, a type, and a set of properties.
- *User* – This interface (which extends Role) is used to represent any entity which may have credentials associated with it. These credentials can be used to authenticate an initiator.
- *Group* – This interface (which extends User) is used to contain an aggregation of named Role objects (Group or User objects).
- *Authorization* – This interface encapsulates an authorization context on which bundles can base authorization decisions.
- *UserAdminEvent* – This class is used to represent a role change event.
- *UserAdminListener* – This interface provides a listener for events of type UserAdminEvent that can be registered as a service.
- *UserAdminPermission* – This permission is needed to configure and access the roles managed by a User Admin service.

Figure 37   *User Admin Service*, org.osgi.service.useradmin

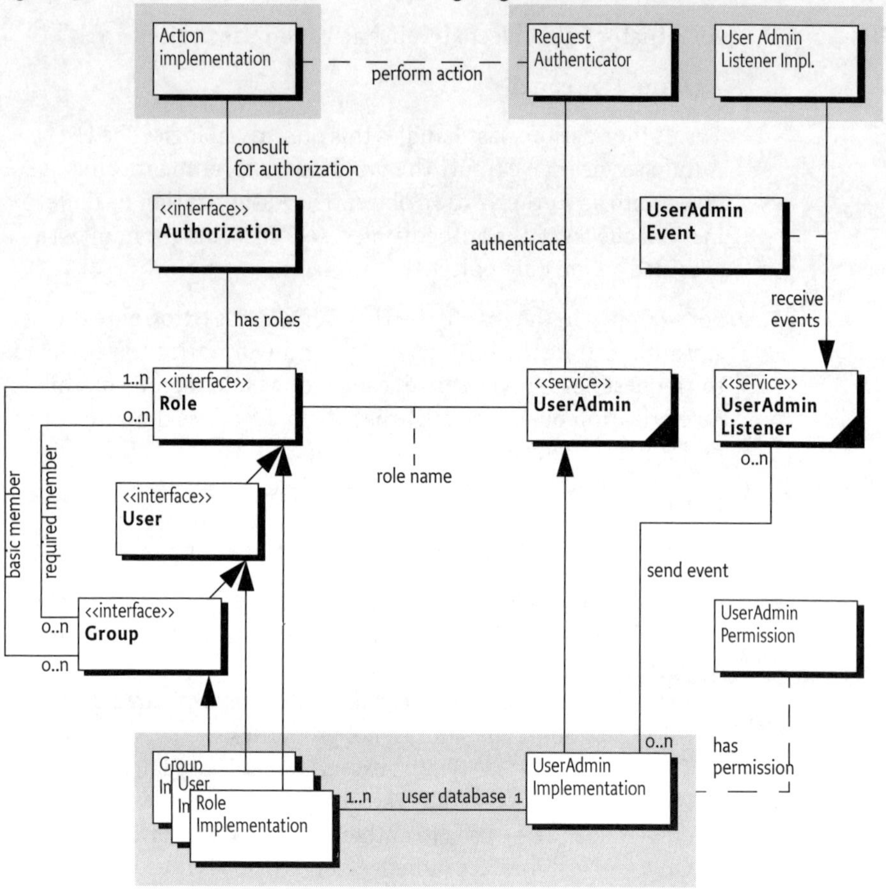

12.2   Authentication

The authentication phase determines if the initiator is genuinely
the one it says it is. Mechanisms to authenticate always need some
information related to the user or the OSGi environment to authenticate an external user. This information can be:

- A secret known only to the initiator.
- Knowledge about cards that can generate a unique token.
- Public information like certificates of trusted signers.
- Information about the user that can be measured in a trusted
  way.
- Other specific information.

12.2.1   Repository

The User Admin service offers a repository of Role objects. Each Role object has a unique name and a set of properties that are readable by anyone, and are changeable when the changer has the UserAdmin-Permission. Additionally, User objects, a sub-interface of Role, also have a set of private protected properties called credentials. Credentials are an extra set of properties that are used to authenticate users and that are protected by UserAdminPermission.

Properties are accessed with the Role.getProperties() method and credentials with the User.getCredentials()method. Both methods return a Dictionary object containing the key/value pairs. The keys are String objects and the values of this Dictionary object are limited to String or byte[ ] objects.

This specification does not define any standard keys for the properties or credentials. The keys depend on the implementation of the authentication mechanism, and are not formally defined by the OSGi.

The repository can be searched for objects that have a unique property (key/value pair) with the method UserAdmin.getUser(String, String). This makes it easy to find a specific user related to a specific authentication mechanism. For example, a secure card mechanism that generates unique tokens could have a serial number identifying the user. The owner of the card could be found with the method

```
User owner = useradmin.getUser(
    "secure-card-serial", "132456712-1212").
```

If multiple User objects have the same property (key *and* value), a null is returned.

There is a convenience method to verify that a user has a credential without actually getting the credential. This is the User.has-Credential(String, Object) method.

Access to credentials is protected on a name basis by UserAdmin-Permission. Because properties are readable to anyone with access to a User object, UserAdminPermission only protects change access to properties.

12.2.2   Basic Authentication

Authentication algorithms that perform validation are wide spread. The following example shows a very simple authentication algorithm based on passwords.

The vendor of the authentication bundle uses the property "com.acme.basic-id" to contain the name of the user as it logs in. This property is used to locate the User object in the repository. Next, the credential "com.acme.password" contains the password and is compared to the entered password. If the password is correct, the User object is returned. In all other cases a SecurityException is thrown.

```
public User authenticate(
      UserAdmin ua, String name, String pwd
) throws SecurityException{
   User user = ua.getUser("com.acme.basicid",
      username);
   if (user == null)
      throw new SecurityException( "No such user" );

   if (!user.hasCredential("com.acme.password", pwd)
      throw new SecurityException(
         "Invalid password" );
   return user;
}
```

12.2.3   Certificates

Authentication based on certificates does not require a shared secret. Instead, a certificate contains a name, a public key, and the signature of one or more signers.

The name in the certificate can be used to locate a User object in the repository. Locating a User object, however, only identifies the initiator and does not authenticate it.

The first step to authenticate the initiator is to verify that it has the private key of the certificate.

Next, the User Admin service must verify that it has a User object with the right property, for example "com.acme.certificate"="Fudd".

The next step is to see if the certificate is signed by a trusted source. The bundle could use a central list of trusted signers and only accept certificates signed by those sources. Alternatively, it could require that the certificate itself is already stored in the repository under a unique key, by storing the certificate as a byte[] in the credentials.

In any case, once the certificate is verified, the associated User object is authenticated.

12.3   Authorization

The User Admin service authorization architecture is a *role based model.* In this model, every action that can be performed by a bundle is associated with a *role.* Such a role is a Group object, (group) from the User Admin service repository. For example, if a servlet could be used to activate the alarm system, there should be a group named AlarmSystemActivation.

The operator can administrate the authorizations by populating the group with User objects (users) and other groups. Groups are used to minimize the amount of administration required. For example, it is easier to create one Administrators group and add it to any administrative roles than to administer all users for each role. Such a group makes it only one action to remove or add a user to act as an administrator.

The authorization decision can now be made in two fundamentally different ways.

An initiator could be allowed to carry out an action (represented by a Group object) if it implied any of the Group object's members. For example, the AlarmSystemActivation Group object contains an Administrators and a Family Group object:

```
Administrators           = { Elmer, Pepe, Bugs }
Family                   = { Elmer, Pepe, Daffy }

AlarmSystemActivation    = { Administrators, Family
}
```

Any of the four members Elmer, Pepe, Daffy or Bugs can activate the alarm system.

Alternatively, an initiator could be allowed to perform an action (represented by a Group object) if it implied *all* of the Group object's members. In this case, using the same AlarmSystemActivation group, only Elmer and Pepe would be authorized to activate the alarm system, since Daffy and Bugs are *not* members of *both* the Administrators and Family Group objects.

The User Admin service supports a combination of both strategies by defining both a set of *basic members* (any) and a set of *required members* (all).

```
Administrators = { Elmer, Pepe, Bugs }
Family         = { Elmer, Pepe, Daffy }

AlarmSystemActivation
```

```
required     = { Administrators }
basic        = { Family }
```

The difference is made when Role objects are added to the Group object. To add a basic member, use the Group.addMember(Role) method. To add a required member, use the Group.addRequired-Member(Role) method.

Basic members *scope* the set of members that can get access, and required members reduce this set by requiring the initiator to *imply* each required member.

A User object implies a Group object if it implies:

- *All* of the Group's required members, and
- At *least* one of the Group's basic members

A User object always implies itself.

If only required members are used to qualify the implication, then the standard user user.anyone can be obtained from the User Admin service and added to the Group object. This Role object is implied by anybody and therefore does not affect the required members.

12.3.1   The Authorization Object

The complexity of the authorization is hidden in an Authorization class. Normally, the authenticator should retrieve an Authorization object from the User Admin service by passing the authenticated User object as argument. This Authorization object is then passed to the bundle that performs the action. This bundle checks the authorization with the Authorization.hasRole(String) method. The performing bundle must pass the name of the action as an argument. The Authorization object checks if the authenticated user implies the Role object, specifically a Group object, with the given name.

```java
public void activateAlarm(Authorization auth) {
    if ( auth.hasRole( "AlarmSystemActivation" ) ) {
        // activate the alarm

        ...

    }
    else throw new SecurityException(
        "Not authorized to activate alarm" );
}
```

12.3.2   The Http Service and Authorization

The Http Service is normally an important entry point of the OSGi environment. It has, therefore, played an important role in the development of the User Admin service development.

The Http Service delegates the authentication to the HttpContext object, which is given as an argument when the servlet is registered. This object should use the User Admin service to authenticate the user. After having obtained a User object, the HttpContext object should create an Authorization object with UserAdmin.getAuthorization(User). This Authorization object captures all the roles of the User object.

The Http Service uses the servlet API to invoke requests on a servlet. This existing API has obviously no place for an OSGi Authorization object. Therefore, the Authorization object is added as an attribute to the javax.servlet.ServletRequest object with the name:

```
org.osgi.service.http.HttpContext.AUTHORIZATION
= ("org.osgi.service.useradmin.authorization")
```

A servlet can retrieve the value of this attribute by calling javax.servlet.ServletRequest.getAttribute(HttpContext.AUTHORIZATION).

12.3.3   Authorization Example

This section demonstrates a possible use of the User Admin service. The service has a flexible model, and many other schemes are possible.

Assume an operator installs an OSGi environment. Bundles in this environment have defined the following action groups:

```
AlarmSystemControl
InternetAccess
TemperatureControl
PhotoAlbumEdit
PhotoAlbumView
PortForwarding
```

Installing and uninstalling bundles could potentially extend this set. Therefore, the operators also defined a number of groups that can be used to contain the different types of system users.

```
Administrators
Buddies
Children
Adults
Residents
```

In a particular instance, the operator installs it in a household with the following residents and buddies:

```
Residents:      Elmer, Fudd, Marvin, Pepe
Buddies:        Daffy, Foghorn
```

First, the residents and acquaintances are assigned to the system user groups. Second, the user groups need to be assigned to the action groups.

The following tables show how the groups could be assigned.

Groups	Elmer	Fudd	Marvin	Pepe	Daffy	Foghorn
Residents	Basic	Basic	Basic	Basic	-	-
Buddies	-	-	-	-	Basic	Basic
Children	-	-	Basic	Basic	-	-
Adults	Basic	Basic	-	-	-	-
Administrators	Basic	-	-	-	-	-

Table 17   *Example Groups with Basic and Required Members*

Groups	Residents	Buddies	Children	Adults	Admin
AlarmSystem-Control	Basic	-	-	-	Required
InternetAccess	Basic	-	-	Required	-
Temperature-Control	Basic	-	-	Required	-
PhotoAlbumEdit	Basic	-	Basic	Basic	-
PhotoAlbum-View	Basic	Basic	-	-	-
PortForwarding	Basic	-	-	-	Required

Table 18   *Example Action Groups with their Basic and Required Members*

12.4   Repository Maintenance

The UserAdmin interface is a straightforward API to maintain a repository of User and Group objects. It contains methods to create new Group and User objects with the createRole(String,int) method. The method is prepared so that the same signature can be used to create new types of roles in the future. The interface also contains a method to remove a Role object.

The existing configuration can be obtained with methods that lists all Role objects using a filter argument. This filter, with the same syntax as the Framework filter, must only return the Role objects where the filter matches the properties.

Several utility methods simplify getting User objects depending on their properties.

12.5 User Admin Events

Changes in the User Admin service can be found out in real time. Each User Admin service implementation must send a UserAdmin-Event object to any service in the Framework service registry that is registered under the UserAdminListener interface.

This procedure is called the white board approach and is demonstrated in the following code sample.

```
class Listener implements UserAdminListener {
   public void roleChanged( UserAdminEvent event ) {
      ...
   }
}
public class MyActivator
   implements BundleActivator {
   public void start( BundleContext conext ) {
      context.registerService( new Listener() );
   }
   public void stop( BundleContext context ) {}
}
```

It is not necessary to unregister the listener. When the bundle is stopped, the Framework must automatically unregister it. Once registered, all changes to the role repository must be notified to the UserAdminListener object.

12.6 Security

The User Admin service is related to the security model of the OSGi environment, but is complementary to the [37] *The Java Security Architecture for JDK 1.2*. The final permission of most code should be the conjunction of the Java 2 Permissions (which are based on the code that is executing) and the User Admin service authorization (which is based on the user for whom the code runs).

12.6.1 UserAdminPermission

The User Admin service defines the UserAdminPermission class that can be used to restrict bundles in accessing the credentials. This permission class has the following actions:

- changeProperty – This permission is required to modify properties. The name of the permission is the prefix of the property name.
- changeCredential – This action permits changing credentials. The name of the permission is the prefix of the name of the credential.
- getCredential – This action permits getting credentials. The name of the permission is the prefix of the credential.

If the name of the permission is "admin", it allows the owner to administer the repository. No action is associated with the permission in that case.

Otherwise, the permission name is used to match the property name. This name may end with a ".*" string to indicate a wildcard. For example com.acme.* matches com.acme.fudd.elmer and com.acme.bugs.

12.7 Relation to JAAS

The Java Authorization and Authentication Service (JAAS) seems at first sight a very suitable model for user administration. The OSGi, however, decided to develop an independent User Admin service because JAAS was not deemed applicable. The reasons were dependency on J2SE version 1.3 ("JDK 1.3") and existing mechanisms in the previous OSGi Service Gateway 1.0 specification.

12.7.1 JDK 1.3 Dependencies

The authorization component of JAAS relies on the java.security.DomainCombiner interface, which provides a means to dynamically update the ProtectionDomain objects affiliated with an AccessControlContext object.

This interface was added in JDK 1.3. In the context of JAAS, the SubjectDomainCombiner object, which implements the DomainCombiner interface, is used to update ProtectionDomain objects, whose permissions depend on where code came from and who signed it, with permissions based on who is running the code.

Leveraging JAAS would have meant that user-based access control on the OSGi environment was only available with JDK 1.3, which was not deemed acceptable.

12.7.2 Existing OSGi Mechanism

JAAS provides a pluggable authentication architecture, which enables applications and their underlying authentication services to remain independent from each other.

The Http Service already provides a similar feature by allowing servlet and resource registrations to be supported by an HttpContext object, which uses a callback mechanism to perform any required authentication checks before granting access to the servlet or resource. This way, the registering bundle has complete control – on a per-servlet and per-resource basis – over which authentication protocol to use, how the credentials presented by the remote requestor are to be validated, and who should be granted access to the servlet or resource.

12.7.3   Future Road Map

In the future, the main barrier of 1.3 compatibility will be removed. In that case JAAS could be implemented on OSGi environment. At that time, the User Admin service will still be needed and will provide complementary services in the following ways:

- The authorization component relies on group membership information to be stored and managed outside JAAS. JAAS does not manage persistent information, so the User Admin service can be a provider of group information when principals are assigned to a Subject object.
- The authorization component allows for credentials to be collected and verified, but again, a repository is needed to actually validate the credentials.

In the future, the User Admin service can act as the back-end database to JAAS. The only aspect JAAS will remove from the User Admin service is the need for the Authorization interface.

12.8   org.osgi.service.useradmin

The OSGi User Admin service Package. Specification Version 1.0.

Bundles wishing to use this package must list the package in the Import-Package header of the bundle's manifest. For example:

```
Import-Package: org.osgi.service.useradmin;
specification-version=1.0
```

12.8.1   Authorization

public interface Authorization

The Authorization interface encapsulates an authorization context on which bundles can base authorization decisions, where appropriate.

Class Summary

Interfaces

Authorization	The Authorization interface encapsulates an authorization context on which bundles can base authorization decisions, where appropriate.
Group	A named grouping of roles (Role objects).
Role	The base interface for Role objects managed by the User Admin service.
User	A User role managed by a User Admin service.
UserAdmin	This interface is used to manage a database of named Role objects, which can be used for authentication and authorization purposes.
UserAdminListener	Listener for UserAdminEvents.

Classes

UserAdminEvent	Role change event.
UserAdminPermission	Permission to configure and access the Role objects managed by a User Admin service.

Bundles associate the privilege to access restricted resources or operations with roles. Before granting access to a restricted resource or operation, a bundle will check if the Authorization object passed to it possesses the required role, by calling its has Role method.

Authorization contexts are instantiated by calling the getAuthorization(User) method.

Trusting Authorization objects

There are no restrictions regarding the creation of Authorization objects. Hence, a service must only accept Authorization objects from bundles that has been authorized to use the service using code based (or Java 2) permissions.

In some cases it is useful to use Service Permission to do the code based access control. A service basing user access control on Authorization objects passed to it, will then require that a calling bundle has the Service Permission to get the service in question. This is the most convenient way. The OSGi environment will do the code based permission check when the calling bundle attempts to get the service from the service registry.

Example: A servlet using a service on a user's behalf. The bundle with the servlet must be given the ServicePermission to get the Http Service.

However, in some cases the code based permission checks need to be more fine-grained. A service might allow all bundles to get it, but require certain code based permissions for some of its methods.

Example: A servlet using a service on a user's behalf, where some service functionality is open to anyone, and some is restricted by code based permissions. When a restricted method is called (e.g., one handing over an Authorization object), the service explicitly checks that the calling bundle has permission to make the call.

12.8.1.1 Methods

public java.lang.String **getName**()

Gets the name of the User that this Authorization context was created for.

Returns: The name of the User object that this Authorization context was created for, or null if no user was specified when this Authorization context was created.

public java.lang.String[] **getRoles**()

Gets the names of all roles encapsulated by this Authorization context.

Returns: The names of all roles encapsulated by this Authorization context, or null if no roles are in the context. The predefined role user.anyone will not be included in this list.

public boolean **hasRole**(java.lang.String name)

Checks if the role with the specified name is implied by this Authorization context.

Bundles must define globally unique role names that are associated with the privilege of accessing restricted resources or operations. Operators will grant users access to these resources, by creating a Group object for each role and adding User objects to it.

Parameters: name - The name of the role to check for.

Returns: true if this Authorization context implies the specified role, otherwise false.

12.8.2   Group

public interface Group extends User

All Superinter-faces: Role, User

A named grouping of roles (Role objects).

Whether or not a given Authorization context implies a Group object depends on the members of that Group object.

A Group object can have two kinds of members: *basic* and *required*. A Group object is implied by an Authorization context if all of its required members are implied and at least one of its basic members is implied.

A Group object must contain at least one basic member in order to be implied. In other words, a Group object without any basic member roles is never implied by any Authorization context.

A User object always implies itself.

No loop detection is performed when adding members to Group objects, which means that it is possible to create circular implications. Loop detection is instead done when roles are checked. The semantics is that if a role depends on itself (i.e., there is an implication loop), the role is not implied.

The rule that a Group object must have at least one basic member to be implied is motivated by the following example:

```
group foo
   required members: marketing
   basic members: alice, bob
```

Privileged operations that require membership in "foo" can be performed only by "alice" and "bob", who are in marketing.

If "alice" and "bob" ever transfer to a different department, anybody in marketing will be able to assume the "foo" role, which certainly must be prevented. Requiring that "foo" (or any Group object for that matter) must have at least one basic member accomplishes that.

However, this would make it impossible for a Group object to be implied by just its required members. An example where this implication might be useful is the following declaration: "Any citizen who is an adult is allowed to vote." An intuitive configuration of "voter" would be:

```
group voter
   required members: citizen, adult
      basic members:
```

However, according to the above rule, the "voter" role could never be assumed by anybody, since it lacks any basic members. In order to address this issue a predefined role named "user.anyone" can be specified, which is always implied. The desired implication of the "voter" group can then be achieved by specifying "user.anyone" as its basic member, as follows:

```
group voter
   required members: citizen, adult
      basic members: user.anyone
```

12.8.2.1   **Methods**

public boolean **addMember**(Role role)
   throws SecurityException

Adds the specified Role object as a basic member to this Group object.

Parameters: role - The role to add as a basic member.

Returns: true if the given role could be added as a basic member, and false if this Group object already contains a Role object whose name matches that of the specified role.

Throws: SecurityException - If a security manager exists and the caller does not have the UserAdminPermission with name admin.

public boolean **addRequiredMember**(Role role)
   throws SecurityException

Adds the specified Role object as a required member to this Group object.

Parameters: role - The Role object to add as a required member.

Returns: true if the given Role object could be added as a required member, and false if this Group object already contains a Role object whose name matches that of the specified role.

Throws: SecurityException - If a security manager exists and the caller does not have the UserAdminPermission with name admin.

public Role[] **getMembers**()

Gets the basic members of this Group object.

Returns: The basic members of this Group object, or null if this Group object does not contain any basic members.

public Role[] **getRequiredMembers**()

Gets the required members of this Group object.

Returns: The required members of this Group object, or null if this Group object does not contain any required members.

public boolean **removeMember**(Role role)
    throws SecurityException

Removes the specified Role object from this Group object.

Parameters: role - The Role object to remove from this Group object.

Returns: true if the Role object could be removed, otherwise false.

Throws: SecurityException - If a security manager exists and the caller does not have the UserAdminPermission with name admin.

12.8.3   Role

public interface Role

All Known Sub-interfaces: Group, User

The base interface for Role objects managed by the User Admin service.

This interface exposes the characteristics shared by all Role classes: a name, a type, and a set of properties.

Properties represent public information about the Role object that can be read by anyone. Specific UserAdminPermission objects are required to change a Role object's properties.

Role object properties are Dictionary objects. Changes to these objects are propagated to the User Admin service and made persistent.

Every User Admin service contains a set of predefined Role objects that are always present and cannot be removed. All predefined Role objects are of type ROLE. This version of the org.osgi.service.useradmin package defines a single predefined role named "user.anyone", which is inherited by any other role. Other predefined roles may be added in the future. Since "user.anyone" is a Role object that has properties associated with it that can be read and modified. Access to these properties and their use is application specific and is controlled using UserAdminPermission in the same way that properties for other Role objects are.

12.8.3.1	**Fields**

public static final int **GROUP**

The type of a Group role.

The value of GROUP is 2.

public static final int **ROLE**

The type of a predefined role.

The value of ROLE is 0.

public static final int **USER**

The type of a User role.

The value of USER is 1.

12.8.3.2	**Methods**

public java.lang.String **getName**()

Returns the name of this role.

Returns: The role's name.

public java.util.Dictionary **getProperties**()

Returns a Dictionary of the (public) properties of this Role object. Any changes to the returned Dictionary will change the properties of this Role object. This will cause a UserAdminEvent object of type ROLE_CHANGED to be broadcast to any UserAdminListener objects.

Only objects of type String may be used as property keys, and only objects of type String or byte[] may be used as property values. Any other types will cause an exception of type IllegalArgumentException to be raised.

In order to add, change, or remove a property in the returned Dictionary, a UserAdminPermission named after the property name (or a prefix of it) with action changeProperty is required.

Returns: Dictionary containing the properties of this Role object.

public int **getType**()

Returns the type of this role.

Returns: The role's type.

12.8.4	**User**

public interface User extends Role

All Known Sub-interfaces: Group

All Superinter-faces: Role

A User role managed by a User Admin service.

In this context, the term "user" is not limited to just human beings. Instead, it refers to any entity that may have any number of credentials associated with it that it may use to authenticate itself.

In general, User objects are associated with a specific User Admin service (namely the one that created them), and cannot be used with other User Admin services.

A User object may have credentials (and properties, inherited from the Role class) associated with it. Specific UserAdminPermission objects are required to read or change a User object's credentials.

Credentials are Dictionary objects and have semantics that are similar to the properties in the Role class.

12.8.4.1   **Methods**

public java.util.Dictionary **getCredentials**()

Returns a Dictionary of the credentials of this User object. Any changes to the returned Dictionary object will change the credentials of this User object. This will cause a UserAdminEvent object of type ROLE_CHANGED to be broadcast to any UserAdminListeners objects.

Only objects of type String may be used as credential keys, and only objects of type String or of type byte[] may be used as credential values. Any other types will cause an exception of type Illegal-ArgumentException to be raised.

In order to retrieve a credential from the returned Dictionary object, a UserAdminPermission named after the credential name (or a prefix of it) with action getCredential is required.

In order to add or remove a credential from the returned Dictionary object, a UserAdminPermission named after the credential name (or a prefix of it) with action changeCredential is required.

Returns: Dictionary object containing the credentials of this User object.

public boolean **hasCredential**(java.lang.String key, java.lang.Object
    value)
    throws SecurityException

Checks to see if this User object has a credential with the specified
key set to the specified value.

If the specified credential value is not of type String or byte[], it is
ignored, that is, false is returned (as opposed to an IllegalArgument-
Exception being raised).

Parameters: key - The credential key.

value - The credential value.

Returns: true if this user has the specified credential; false otherwise.

Throws: SecurityException - If a security manager exists and the caller does
                     not have the UserAdminPermission named after the credential key (
                     or a prefix of it) with action getCredential.

12.8.5   UserAdmin

public interface UserAdmin

This interface is used to manage a database of named Role objects,
which can be used for authentication and authorization purposes.

This version of the User Admin service defines two types of Role
objects: "User" and "Group". Each type of role is represented by an
int constant and an interface. The range of positive integers is
reserved for new types of roles that may be added in the future.
When defining proprietary role types, negative constant values
must be used.

Every role has a name and a type.

A User object can be configured with credentials (e.g., a password)
and properties (e.g., a street address, phone number, etc.).

A Group object represents an aggregation of User and Group
objects. In other words, the members of a Group object are roles
themselves.

Every User Admin service manages and maintains its own
namespace of Role objects, in which each Role object has a unique
name.

12.8.5.1   Methods
public Role **createRole**(java.lang.String name, int type)
    throws IllegalArgumentException, SecurityException

Creates a Role object with the given name and of the given type.

If a Role object was created, a UserAdminEvent object of type
ROLE_CREATED is broadcast to any UserAdminListener object.

Parameters:		name - The name of the Role object to create.

type - The type of the Role object to create. Must be either a USER
type or GROUP type.

Returns:		The newly created Role object, or null if a role with the given name
already exists.

Throws:		IllegalArgumentException - if type is invalid.

SecurityException - If a security manager exists and the caller does
not have the UserAdminPermission with name admin.

public Authorization **getAuthorization**(User user)

Creates an Authorization object that encapsulates the specified User
object and the Role objects it possesses. The null user is interpreted
as the anonymous user. The anonymous user represents a user that
has not been authenticated. An Authorization object for an anonymous user will be unnamed, and will only imply groups that
user.anyone implies.

Parameters:		user - The User object to create an Authorization object for, or null
for the anonymous user.

Returns:		the Authorization object for the specified User object.

public Role **getRole**(java.lang.String name)

Gets the Role object with the given name from this User Admin service.

Parameters:		name - The name of the Role object to get.

Returns:		The requested Role object, or null if this User Admin service does not
have a Role object with the given name.

public Role[] **getRoles**(java.lang.String filter)

Gets the Role objects managed by this User Admin service that have
properties matching the specified LDAP filter criteria. See
org.osgi.framework.Filter for a description of the filter syntax. If a
null filter is specified, all Role objects managed by this User Admin
service are returned.

Parameters:		filter - The filter criteria to match.

Returns: The Role objects managed by this User Admin service whose properties match the specified filter criteria, or all Role objects if a null filter is specified. If no roles match the filter, null will be returned.

Throws: InvalidSyntaxException

public User **getUser**(java.lang.String key, java.lang.String value)

Gets the user with the given property key-value pair from the User Admin service database. This is a convenience method for retrieving a User object based on a property for which every User object is supposed to have a unique value (within the scope of this User Admin service), such as for example a X.500 distinguished name.

Parameters: key - The property key to look for.

value - The property value to compare with.

Returns: A matching user, if *exactly* one is found. If zero or more than one matching users are found, null is returned.

public boolean **removeRole**(java.lang.String name)
    throws SecurityException

Removes the Role object with the given name from this User Admin service.

If the Role object was removed, a User Admin Event object of type ROLE_REMOVED is broadcast to any User Admin Listener object.

Parameters: name - The name of the Role object to remove.

Returns: true If a Role object with the given name is present in this User Admin service and could be removed, otherwise false.

Throws: SecurityException - If a security manager exists and the caller does not have the User Admin Permission with name admin.

12.8.6   UserAdminEvent

public class UserAdminEvent

Role change event.

User Admin Event objects are delivered asynchronously to any User-Admin Listener objects when a change occurs in any of the Role objects managed by a User Admin service.

A type code is used to identify the event. The following event types are defined: ROLE_CREATED type, ROLE_CHANGED type, and ROLE_REMOVED type. Additional event types may be defined in the future.

See Also:	UserAdmin, UserAdminListener

12.8.6.1   **Fields**

public static final int **ROLE_CHANGED**

A Role object has been modified.

The value of ROLE_CHANGED is 0x00000002.

public static final int **ROLE_CREATED**

A Role object has been created.

The value of ROLE_CREATED is 0x00000001.

public static final int **ROLE_REMOVED**

A Role object has been removed.

The value of ROLE_REMOVED is 0x00000004.

12.8.6.2   **Constructors**

public **UserAdminEvent**(ServiceReference ref, int type, Role role)

Constructs a UserAdminEvent object from the given Service-Reference object, event type, and Role object.

Parameters: ref - The ServiceReference object of the User Admin service that generated this event.

type - The event type.

role - The Role object on which this event occurred.

12.8.6.3   **Methods**

public Role **getRole**()

Gets the Role object this event was generated for.

Returns: The Role object this event was generated for.

public ServiceReference **getServiceReference**()

Gets the ServiceReference object of the User Admin service that generated this event.

Returns: The User Admin service's ServiceReference object.

public int **getType**()

Returns the type of this event.

The type values are ROLE_CREATED type, ROLE_CHANGED type, and ROLE_REMOVED type.

Returns: The event type.

12.8.7   UserAdminListener

public interface UserAdminListener

Listener for UserAdminEvents.

UserAdminListener objects are registered with the Framework service registry and notified with a UserAdminEvent object when a Role object has been created, removed, or modified.

UserAdminListener objects can further inspect the received UserAdminEvent object to determine its type, the Role object it occurred on, and the User Admin service that generated it.

See Also: UserAdmin, UserAdminEvent

12.8.7.1   Methods
public void **roleChanged**(UserAdminEvent event)

Receives notification that a Role object has been created, removed, or modified.

Parameters: event - The UserAdminEvent object.

12.8.8   UserAdminPermission

public final class UserAdminPermission extends
    java.security.BasicPermission

All Implement-
ed Interfaces: java.security.Guard, java.io.Serializable

Permission to configure and access the Role objects managed by a User Admin service.

This class represents access to the Role objects managed by a User Admin service and their properties and credentials (in the case of User objects).

The permission name is the name (or name prefix) of a property or credential. The naming convention follows the hierarchical property naming convention. Also, an asterisk may appear at the end of the name, following a ".", or by itself, to signify a wildcard match. For example: "org.osgi.security.protocol.*" or "*" is valid, but "*protocol" or "a*b" are not valid.

The UserAdminPermission with the reserved name "admin" represents the permission required for creating and removing Role objects in the User Admin service, as well as adding and removing members in a Group object. This UserAdminPermission does not have any actions associated with it.

The actions to be granted are passed to the constructor in a string containing a list of one or more comma-separated keywords. The possible keywords are: "changeProperty", "changeCredential", and "getCredential". Their meaning is defined as follows:

```
  action: "changeProperty"
   Permission to change (i.e., add and remove) Role
object properties
   whose names start with the name argument specified
in the
   constructor.
  action: "changeCredential"
   Permission to change (i.e., add and remove) User
object credentials
   whose names start with the name argument specified
in the
   constructor.
  action: "getCredential"
   Permission to retrieve and check for the existence
of User object
   credentials whose names start with the name
argument
   specified in the constructor.
```

The action string is converted to lowercase before processing.

Following is a Java 2 style policy entry which grants a user administration bundle a number of UserAdminPermission object:

```
  grant codeBase "${jars}useradmin_console.jar" {
     permission
  org.osgi.service.useradmin.UserAdminPermission
  "admin";
     permission
  org.osgi.service.useradmin.UserAdminPermission
                "com.foo.*", "changeProperty,
  getCredential,changeCredential";
     permission
  org.osgi.service.useradmin.UserAdminPermission
                "user.*", "changeProperty,
  changeCredential";
  };
```

The first permission statement grants the bundle the permission to perform any User Admin service operations of type "admin", that is, create and remove roles and configure Group objects.

The second permission statement grants the bundle the permission to change any properties as well as get and change any credentials whose names start with com.foo..

The third permission statement grants the bundle the permission to change any properties and credentials whose names start with user.. This means that the bundle is allowed to change, but not retrieve any credentials with the given prefix.

The following policy entry empowers the Http Service bundle to perform user authentication:

```
grant codeBase "${jars}http.jar" {
   permission
 org.osgi.service.useradmin.UserAdminPermission
              "user.password", "getCredential";
   };
```

The permission statement grants the Http Service bundle the permission to validate any password credentials (for authentication purposes), but the bundle is not allowed to change any properties or credentials.

12.8.8.1   **Fields**

public static final java.lang.String **ADMIN**

The permission name "admin".

public static final java.lang.String **CHANGE_CREDENTIAL**

The action string "changeCredential".

public static final java.lang.String **CHANGE_PROPERTY**

The action string "changeProperty".

public static final java.lang.String **GET_CREDENTIAL**

The action string "getCredential".

12.8.8.2   **Constructors**

public **UserAdminPermission**(java.lang.String name,
    java.lang.String actions)
    throws IllegalArgumentException

Creates a new User Admin Permission with the specified name and
actions. name is either the reserved string "admin" or the name of a
credential or property, and actions contains a comma-separated list
of the actions granted on the specified name. Valid actions are
"changeProperty", "changeCredential", and "getCredential".

Parameters: name - the name of this User Admin Permission

actions - the action string.

Throws: IllegalArgumentException - If name equals "admin" and actions are
specified.

12.8.8.3   **Methods**

public boolean **equals**(java.lang.Object obj)

Checks two User Admin Permission objects for equality. Checks that
obj is a User Admin Permission, and has the same name and actions
as this object.

Overrides: java.security.BasicPermission.equals(java.lang.Object) in class
java.security.BasicPermission

Parameters: obj - the object to be compared for equality with this object.

Returns: true if obj is a User Admin Permission object, and has the same name
and actions as this User Admin Permission object.

public java.lang.String **getActions**()

Returns the canonical string representation of the actions, separated
by comma.

Overrides: java.security.BasicPermission.getActions() in class java.secu-
rity.BasicPermission

Returns: the canonical string representation of the actions.

public int **hashCode**()

Returns the hash code of this User Admin Permission object.

Overrides: java.security.BasicPermission.hashCode() in class java.secu-
rity.BasicPermission

public boolean **implies**(java.security.Permission p)

Checks if this User Admin Permission object "implies" the specified permission.

More specifically, this method returns true if:

- *p* is an instanceof User Admin Permission,
- *p*'s actions are a proper subset of this object's actions, and
- *p*'s name is implied by this object's name. For example, "java.*" implies "java.home".

Overrides: java.security.BasicPermission.implies(java.security.Permission) in class java.security.BasicPermission

Parameters: p - the permission to check against.

Returns: true if the specified permission is implied by this object; false otherwise.

public java.security.PermissionCollection
  newPermissionCollection()

Returns a new Permission Collection object for storing User Admin-Permission objects.

Overrides: java.security.BasicPermission.newPermissionCollection() in class java.security.BasicPermission

Returns: a new Permission Collection object suitable for storing User Admin-Permission objects.

public java.lang.String **toString**()

Returns a string describing this User Admin Permission. The convention is to specify the class name, the permission name, and the actions in the following format: '(org.osgi.service.useradmin.User-AdminPermission "name" "actions")'.

Overrides: java.security.Permission.toString() in class java.security.Permission

Returns: information about this Permission object.

12.9   References

[37] *The Java Security Architecture for JDK 1.2*
Version 1.0, Sun Microsystems, October 1998
http://java.sun.com/products/jdk/1.2/docs/guide/security/spec/security-spec.doc.html

[38] *Java Authentication and Authorization Service*
      http://java.sun.com/products/jaas/

Index

A

B

C

capturing
events 51
CHANGE_CREDENTIAL
of org.osgi.service.useradmin.UserAdminPermission 337
CHANGE_PROPERTY
of org.osgi.service.useradmin.UserAdminPermission 337
CHARACTER
of org.osgi.service.metatype.AttributeDefinition 280
check
permission 52
childrenNames()
of org.osgi.service.prefs.Preferences 298
clear()
of org.osgi.service.prefs.Preferences 299
close()
of org.osgi.util.tracker.ServiceTracker 144
CM_TARGET
of org.osgi.service.cm.ConfigurationPlugin 264
Configurable
of org.osgi.framework 91
Configuration
of org.osgi.service.cm 255
configuration
managed service 234
properties 232
configuration data
modifying 246
configuration object
accessing 244
deletion 245
getting 248
location binding 231
managed service 243

managed service factory 244
updating 245
configuration plugin
forcing a call-back of 248
modifying data 248
configuration plugin object
calling order 248
registration 252
ConfigurationAdmin
of org.osgi.service.cm 259
ConfigurationException
of org.osgi.service.cm 263
ConfigurationException(String, String)
of org.osgi.service.cm.ConfigurationException 263
ConfigurationPlugin
of org.osgi.service.cm 263
Constants
of org.osgi.framework 92
of org.osgi.service.device 215
context
of org.osgi.util.tracker.ServiceTracker 142
createDefaultHttpContext()
of org.osgi.service.http.HttpService 180
createFactoryConfiguration(String)
of org.osgi.service.cm.ConfigurationAdmin 260
createFactoryConfiguration(String, String)
of org.osgi.service.cm.ConfigurationAdmin 260
createFilter(String)
of org.osgi.framework.BundleContext 79
createRole(String, int)
of org.osgi.service.useradmin.UserAdmin 331
customizing
a service tracker 140

D

default
HTTP context object 170
default HTTP context
sharing 171
delete()
of org.osgi.service.cm.Configuration 256
deleted(String)
of org.osgi.service.cm.ManagedServiceFactory 269
deletion
configuration object 245
managed service 237
managed service factory 241
dependencies
BundleClassPath 24
resolving 30
Device
of org.osgi.service.device 216
device
representation 188
device manager
optimizations 211
starting 206
device service
attachment to 190
registration of 189, 213

unregistration 190
DEVICE_CATEGORY
of org.osgi.service.device.Constants 216
DEVICE_DESCRIPTION
of org.osgi.service.device.Constants 216
DEVICE_SERIAL
of org.osgi.service.device.Constants 216
DOUBLE
of org.osgi.service.metatype.AttributeDefinition 280
Driver
of org.osgi.service.device 217
driver bundle
reclamation 212
updates 212
driver service
attachment of 190
registration of 199, 213
unregistration 200
DRIVER_ID
of org.osgi.service.device.Constants 216
DriverLocator
of org.osgi.service.device 219
DriverSelector
of org.osgi.service.device 220

E

equals(Object)
of org.osgi.framework.AdminPermission 64
of org.osgi.framework.Filter 101

of org.osgi.framework.PackagePermission 106
of org.osgi.framework.ServicePermission 112
of org.osgi.service.permissionadmin.PermissionInfo 133

H

handleSecurity(HttpServletRequest, HttpServletResponse)
of org.osgi.service.http.HttpContext 178
hasCredential(String, Object)
of org.osgi.service.useradmin.User 331
hashCode()
of org.osgi.framework.Filter 101
of org.osgi.framework.PackagePermission 106
of org.osgi.framework.ServicePermission 112
of org.osgi.service.permissionadmin.PermissionInfo 135
of org.osgi.service.useradmin.UserAdminPermission 338
hasPermission(Object)
of org.osgi.framework.Bundle 69
hasRole(String)
of org.osgi.service.useradmin.Authorization 325
headers
authentication 174
HTTP context object
default 170
using 170
HttpContext
of org.osgi.service.http 177
HttpService
of org.osgi.service.http 180

I

implies(Permission)
of org.osgi.framework.AdminPermission 64
of org.osgi.framework.PackagePermission 107
of org.osgi.framework.ServicePermission 113
of org.osgi.service.useradmin.UserAdminPermission 339
IMPORT
of org.osgi.framework.PackagePermission 105
IMPORT_PACKAGE
of org.osgi.framework.Constants 97
IMPORT_SERVICE
of org.osgi.framework.Constants 97
importing
packages 22
services 47
installBundle(String)
of org.osgi.framework.BundleContext 84
installBundle(String, InputStream)
of org.osgi.framework.BundleContext 85
INSTALLED
of org.osgi.framework.Bundle 65
of org.osgi.framework.BundleEvent 89
installing
bundles 30
INTEGER
of org.osgi.service.metatype.AttributeDefinition 281
InvalidSyntaxException
of org.osgi.framework 104
InvalidSyntaxException(String, String)
of org.osgi.framework.InvalidSyntaxException 104
isRemovalPending()
of org.osgi.service.packageadmin.ExportedPackage 123

K

keys()
of org.osgi.service.prefs.Preferences 303

L

Legal terms v
listConfigurations(String)
of org.osgi.service.cm.ConfigurationAdmin 262
listeners
types of 50
loadDriver(String)
of org.osgi.service.device.DriverLocator 219
loading
native language code libraries 25
log entries
retrieving 152
log entry objects
retrieving 153
log(int, String)
of org.osgi.service.log.LogService 160
log(int, String, Throwable)
of org.osgi.service.log.LogService 161
log(ServiceReference, int, String)
of org.osgi.service.log.LogService 161
log(ServiceReference, int, String, Throwable)
of org.osgi.service.log.LogService 161
LOG_DEBUG
of org.osgi.service.log.LogService 160
LOG_ERROR
of org.osgi.service.log.LogService 160
LOG_INFO
of org.osgi.service.log.LogService 160
LOG_WARNING
of org.osgi.service.log.LogService 160
LogEntry
of org.osgi.service.log 157
logged(LogEntry)
of org.osgi.service.log.LogListener 158
LogListener
of org.osgi.service.log 158
LogReaderService
of org.osgi.service.log 158
LogService
of org.osgi.service.log 160
LONG
of org.osgi.service.metatype.AttributeDefinition 281

M

managed service
configuration 234
configuration object
creating 243
deletion 237
example 236

managed service factory
configuration object
creating 244
deletion 241
example 241
registration 239
ManagedService
of org.osgi.service.cm 265
ManagedServiceFactory
of org.osgi.service.cm 267
manifest headers
grammar 18
retrieving 18
mapping
framework events 153
HTTP requests to servlet and resource registrations 169
Match
of org.osgi.service.device 220
match(Dictionary)
of org.osgi.framework.Filter 102
match(ServiceReference)
of org.osgi.framework.Filter 102

of org.osgi.service.device.Driver 218
MATCH_NONE
of org.osgi.service.device.Device 217
message
error condition 151
error severity 152
log level 152
logging methods 150
MetaTypeProvider
of org.osgi.service.metatype 284
metatypes
using 250
MIME
types 171
returning 172
MODIFIED
of org.osgi.framework.ServiceEvent 108
modifiedService(ServiceReference, Object)
of org.osgi.util.tracker.ServiceTracker 146
of org.osgi.util.tracker.ServiceTrackerCustomizer 148
modifyConfiguration(ServiceReference, Dictionary)
of org.osgi.service.cm.ConfigurationPlugin 265

N

name()
of org.osgi.service.prefs.Preferences 304
NamespaceException
of org.osgi.service.http 183
NamespaceException(String)
of org.osgi.service.http.NamespaceException 183
NamespaceException(String, Throwable)
of org.osgi.service.http.NamespaceException 183
native language code
algorithm 26
native language code libraries
loading 25

newPermissionCollection()
of org.osgi.framework.AdminPermission 64
of org.osgi.framework.PackagePermission 107
of org.osgi.framework.ServicePermission 113
of org.osgi.service.useradmin.UserAdminPermission 339
node(String)
of org.osgi.service.prefs.Preferences 304
nodeExists(String)
of org.osgi.service.prefs.Preferences 304
noDriverFound()
of org.osgi.service.device.Device 217

O

OBJECTCLASS
of org.osgi.framework.Constants 97
ObjectClassDefinition
of org.osgi.service.metatype 284
obtaining
services 43
open()
of org.osgi.util.tracker.ServiceTracker 146
optimizations
device manager 211
OPTIONAL
of org.osgi.service.metatype.ObjectClassDefinition 284
org.osgi.framework
package 62
org.osgi.service.cm
package 254
org.osgi.service.device

package 215
org.osgi.service.http
package 176
org.osgi.service.log
package 156
org.osgi.service.metatype
package 279
org.osgi.service.packageadmin
package 121
org.osgi.service.permissionadmin
package 130
org.osgi.service.prefs
package 296
org.osgi.service.useradmin
package 323
org.osgi.util.tracker
package 141

P

PACKAGE_SPECIFICATION_VERSION
of org.osgi.framework.Constants 97
PackageAdmin
of org.osgi.service.packageadmin 123
PackagePermission
of org.osgi.framework 105
PackagePermission(String, String)
of org.osgi.framework.PackagePermission 105
packages

exporting 21
importing 22
sharing 20
parent()
of org.osgi.service.prefs.Preferences 305
PermissionAdmin
of org.osgi.service.permissionadmin 130
PermissionInfo
of org.osgi.service.permissionadmin 132

R

S

security
implementing 251
permissions 251
select(ServiceReference, Match[])
of org.osgi.service.device.DriverSelector 220
SELECT_NONE
of org.osgi.service.device.DriverSelector 220
service factories
using 46
service interfaces
accessing 38
service persistent ID 189
service reference objects
getting 38
service tracker
customizing 140
using 139
SERVICE_DESCRIPTION
of org.osgi.framework.Constants 98
SERVICE_ID
of org.osgi.framework.Constants 98
SERVICE_PID
of org.osgi.framework.Constants 98
SERVICE_RANKING
of org.osgi.framework.Constants 98
SERVICE_VENDOR
of org.osgi.framework.Constants 99
serviceChanged(ServiceEvent)
of org.osgi.framework.ServiceListener 111
ServiceEvent
of org.osgi.framework 107
ServiceEvent(int, ServiceReference)
of org.osgi.framework.ServiceEvent 108
ServiceFactory
of org.osgi.framework 109
ServiceListener
of org.osgi.framework 110
ServicePermission
of org.osgi.framework 111
ServicePermission(String, String)
of org.osgi.framework.ServicePermission 111
ServiceReference
of org.osgi.framework 113
ServiceRegistration
of org.osgi.framework 115
services 37
exporting 47
importing 47
obtaining 43
registering 39
releasing 48
returning registered 45
unregistering 48
ServiceTracker
of org.osgi.util.tracker 142
ServiceTracker(BundleContext, Filter, ServiceTracker-Customizer)
of org.osgi.util.tracker.ServiceTracker 143

ServiceTracker(BundleContext, ServiceReference, ServiceTrackerCustomizer)
of org.osgi.util.tracker.ServiceTracker 143
ServiceTracker(BundleContext, String, ServiceTracker-Customizer)
of org.osgi.util.tracker.ServiceTracker 143
ServiceTrackerCustomizer
of org.osgi.util.tracker 147
servlet
registration 164
servlets
registering 164
setBundleLocation(String)
of org.osgi.service.cm.Configuration 257
setDefaultPermissions(PermissionInfo[])
of org.osgi.service.permissionadmin.PermissionAdmin 132
setPermissions(String, PermissionInfo[])
of org.osgi.service.permissionadmin.PermissionAdmin 132
setProperties(Dictionary)
of org.osgi.framework.ServiceRegistration 115
sharing
packages 20
SHORT
of org.osgi.service.metatype.AttributeDefinition 281
size()
of org.osgi.util.tracker.ServiceTracker 147
specifications
device catetogy 190
start()
of org.osgi.framework.Bundle 70
start(BundleContext)
of org.osgi.framework.BundleActivator 76
STARTED
of org.osgi.framework.BundleEvent 89
of org.osgi.framework.FrameworkEvent 103
STARTING
of org.osgi.framework.Bundle 66
starting
bundles 31
device manager 206
stop()
of org.osgi.framework.Bundle 71
stop(BundleContext)
of org.osgi.framework.BundleActivator 76
STOPPED
of org.osgi.framework.BundleEvent 89
STOPPING
of org.osgi.framework.Bundle 66
stopping
bundles 32
STRING
of org.osgi.service.metatype.AttributeDefinition 281
sync()
of org.osgi.service.prefs.Preferences 309
SynchronousBundleListener
of org.osgi.framework 117
SYSTEM_BUNDLE_LOCATION
of org.osgi.framework.Constants 99

T

toString()
of org.osgi.framework.Filter 102
of org.osgi.service.permissionadmin.PermissionInfo 135
of org.osgi.service.useradmin.UserAdminPermission 339
types
MIME 171
permissions 54

U

ungetService(Bundle, ServiceRegistration, Object)
of org.osgi.framework.ServiceFactory 110
ungetService(ServiceReference)
of org.osgi.framework.BundleContext 88
uninstall()
of org.osgi.framework.Bundle 72
UNINSTALLED
of org.osgi.framework.Bundle 66
of org.osgi.framework.BundleEvent 89
uninstalling
bundles 33
unregister()
of org.osgi.framework.ServiceRegistration 116
unregister(String)
of org.osgi.service.http.HttpService 182
UNREGISTERING
of org.osgi.framework.ServiceEvent 108
unregistering
services 48
unregistration
device service 190
driver service 200
update()
of org.osgi.framework.Bundle 73
of org.osgi.service.cm.Configuration 258
update(Dictionary)
of org.osgi.service.cm.Configuration 258
update(InputStream)

of org.osgi.framework.Bundle 75
UPDATED
of org.osgi.framework.BundleEvent 90
updated(Dictionary)
of org.osgi.service.cm.ManagedService 267
updated(String, Dictionary)
of org.osgi.service.cm.ManagedServiceFactory 269
updating
bundles 32
USER
of org.osgi.service.useradmin.Role 329
User
of org.osgi.service.useradmin 330
UserAdmin
of org.osgi.service.useradmin 331
UserAdminEvent
of org.osgi.service.useradmin 333
UserAdminEvent(ServiceReference, int, Role)
of org.osgi.service.useradmin.UserAdminEvent 334
UserAdminListener
of org.osgi.service.useradmin 335
UserAdminPermission
of org.osgi.service.useradmin 335
UserAdminPermission(String, String)
of org.osgi.service.useradmin.UserAdminPermission 338
using
service tracker 139

V

validate(String)   of org.osgi.service.metatype.AttributeDefinition 283

W

waitForService(long)
of org.osgi.util.tracker.ServiceTracker 147